AF602051

Commercial Plant Breeding Objective

The Authors

Dr. Phundan Singh hails from an agricultural family of Meerut District of Uttar Pradesh. He graduated in Agriculture from Agra University, Agra and obtained his Master and Doctorate degrees from Kanpur University, Kanpur. He has throughout a brilliant academic record. His area of specialization is Plant Breeding and Genetics.

Dr. P. Singh has over 40 year experience in the field of Plant Breeding and Genetics. He has 300 publications to his credit. He has participated and presented 80 research papers in various National and International Seminars, Symposia and Conferences. He has authored more than 60 books, 18 technical bulletins and contributed 14 chapters in various books.

Dr P. Singh is Expert member of Selection Committees and M. Sc. and Ph. D. Examiner in different Agricultural Universities. The Institute received the best Annual Report Award in 2003 when he was the Director. He was awarded Life time achievement award by Cotton Research and Development Association in 2018. He has visited several countries such as Russia, Belarus, Ukraine, Crimea, Uzbekistan, Canada, USA, England, Austria and Germany.

Dr. (Mrs.) Monika Singh belongs to a farming family of Damoh District of Madhya Pradesh. She obtained her B. Sc. (Ag.) and M. Sc. (Ag.) degrees from College of Agriculture, Indore and Ph. D. Degree from JNKVV, Jabalpur. Her specialization is in Plant Breeding and Genetics. Dr. Monika Singh has 10 Research papers besides 1 popular article and three abstract to her credit. She has about four years of teaching and research experience. She has participated and presented 4 research papers in various National and International Symposia and Conferences.

Dr. (Mrs.) Mridula Billore hails from Jabalpur city of Madhya Pradesh. She obtained her M. Sc. and Ph. D. Degrees from JNKVV, Jabalpur in Genetics and Plant Breeding. Dr. Billore has more than 30 year experience of teaching, research and extension. Presently, she is the Dean Faculty at the RVSKVV, Gwalior. She has published more than 70 research papers in various journals. She has authored 2 book and guided more than 50 M. Sc. students. She has evolved a total of 4 varieties of pulse and maize. She served as member of Academic Council of JNKVV, Jabalpur and Board member in Central Agricultural University, Jhansi. She received best voluntary pulse research centre award of ICAR in 2010.

Commercial Plant Breeding Objective

[As per 5th Deans Committee of ICAR]

— *Authors* —

Dr. Phundan Singh

Former Director
ICAR-Central Institute for Cotton Research,
Nagpur – 440 010, Maharashtra, India

Dr. Monika Singh

Technical Assistant [Plant Breeding and Genetics]
JNKVV, Sugarcane Research Station,
Bohani, Narsinghpur, Madhya Pradesh

Dr. Mridula Billore

Dean Faculty
College of Agriculture
RVSKVV, Gwalior [M. P.]

2021

Daya Publishing House®
A Division of
Astral International Pvt. Ltd.
New Delhi – 110 002

ISBN : 9789390371877 (Int. Edition)

Published by : **Daya Publishing House®**
A Division of
Astral International Pvt. Ltd.
– ISO 9001:2015 Certified Company –
4736/23, Ansari Road, Darya Ganj,
New Delhi-110 002
Ph. 011-43549197, 23278134
E-mail: info@astralint.com
Website: www.astralint.com

Dedicated to following Family members of Senior Author

Dr. (Mrs.) Saksham Singh [Daughter-in-law]

Engineer (Miss) Subhika Singh [Granddaughter]

Mr. Rishabh Singh [Grandson]

Preface

Plant breeding plays an important role in genetic improvement of crop plants in relation to their economic use for human being. As per 5th Deans Committee Report of ICAR, a new paper entitled "Commercial Plant Breeding" has been introduced. There are several books of Plant Breeding. However, there is hardly any book which deals the subject in objective manner. The present book has been designed to fill up this gap. In this book few examples related to multiple choice questions, true or false, fill in the blanks and one word answer are cited. Students can frame lot more such questions with the help of their teachers.

This book has been designed to covers the undergraduate syllabus of Commercial Plant Breeding as per ICAR 5th Dean committee report of ICAR. It covers the entire syllabus in one compact volume of 15 chapters. Glossary of technical terms is presented at the end. Some useful information has been appended for general knowledge of students.

The book has been written in a simple language for easy grasping of undergraduate students. Each chapter has been presented point-wise and step by step manner.

The information contained in this book has been gathered from various published sources and Internet websites. Attempts have been made to provide latest information even then some valuable information might have been missed.

The cooperation extended by the wife of senior author Mrs. Jaswanti Singh during preparation of this book is praiseworthy. Hope, this Volume would be useful to the students, researchers, and teachers engaged in the field of Plant Breeding. Constructive suggestions of students and teachers are invited for further improvement of this book.

Phundan Singh

Monika Singh

Mridula Billore

Contents

Syllabus
[As per 5th Deans Committee Report of ICAR]

Types of crops and modes of plant reproduction; Line development and maintenance breeding in self and cross pollinated crops (A/B/R and two line system) for development of hybrids and seed production; Genetic purity test of commercial hybrids. Advances in hybrid seed production of maize, rice, sorghum, pearl millet, castor, sunflower, cotton, pigeon pea, Brassica *etc.*; Quality seed production of vegetable crops under open and protected environments; Alternative strategies for the development of the line and cultivars: haploid inducer, tissue culture techniques and biotechnological tools; IPR issues in commercial plant breeding; DUS testing and registration of varieties under PPV and FR Act; Variety testing, release and notification systems in India; Principles and techniques of seed production; types of seeds; quality testing in self and cross pollinated crops.

Part I

Introductory Chapters

Chapter 1

Modes of Reproduction in Crop Plants

A: DEFINITIONS

Adventive Embryony: Development of embryo from the diploid cells of ovule lying outside the embryosac belonging to either nucellus or integument.

Allogamy: Development of seed by cross pollination.

Anthesis: The process of dehiscence of anthers, the period of pollen distribution.

Apomixis: Development of seed without sexual fusion (fertilization).

Apogamy: Development of embryo either from synergids or antipodal cells of embryosac.

Apospory: Development of another embryosac without reduction from the cell of ovule outside the embryosac and then development of embryo directly from diploid egg cell.

Artificial Vegetative Reproduction: It is achieved by artificial means.

Asexual Reproduction: Multiplication of plants without the fusion of male and female gametes. It is of two type, *viz.* : vegetative reproduction and apomixis.

Autogamy: Development of seed by self-pollination.

Bisexuality: Presence of male and female organs in the same flower.

Chasmogamy: Fertilization after opening of flower.

Cleistogamy: Completion of pollination and fertilization in unopened flower bud.

Dichogamy: Maturation of anthers and stigma of a flower at different time.

Dioecy: Existence of male and female flowers on different plants.

Diploid Parthenogenesis: Development of embryo from diploid egg cell.

Facultative Apomixis: Apomixis in which sexual reproduction occurs in addition to apomixis.

Generative Apospory: Development of new embryosac from archesporium

Haploid Parthenogenesis: Development of embryo from haploid egg cell.

Herkogamy: Hindrance to self-pollination due to presence of hyline membrane around anther.

Heterostyly: Different lengths of styles and filaments in a flower.

Homogamy: Maturation of anthers and stigma of a flower at the same time.

Monoecy: Separate existence of male and female flowers on the same plant in the same inflorescence or different inflorescence.

Natural Vegetative Reproduction: It occurs in nature.

Non-Recurrent Apomixis: When apomictic plants are not obtained from one generation to another.

Obligate Apomixix: Apomixis in which reproduction occurs only by apomixis.

Parthenogenesis: Development of embryo from the egg cell without fertilization.

Recurrent Apomixis: Occurrence of apomixis from one generation to another.

Reproduction: The process by which living organisms give rise to the offspring of similar kind (species).

Somatic Apospory: Development of new embryosac from nucellus or integument.

Sexual Reproduction: Multiplication of plants through fertilized embryos.

Vegetative Reproduction: Multiplication of plants by means of various vegetative plant parts such as stem cuttings and root cuttings.

B: BACKGROUND INFORMATION

1. Reproduction is the process of multiplication of living, beings. In crop plants, the mode of reproduction is of two types, *viz.* (1) sexual reproduction and (2) asexual reproduction.

2. Multiplication of plants through embryos which have developed by fusion of male and female gametes is known as sexual reproduction. All the seed propagating species belong to this group.

3. Sexual reproduction plays an important role in creating and maintaining vast genetic variability in crop plants, which is essential for crop improvement.

4. Sexual reproduction offers opportunities to combine desirable genes together from different sources. Crossing over takes place only during meiosis and leads to new gene combinations.

5. Pollination refers to the process by which pollen grains are transferred from anthers to stigma. Pollination is of two types: *viz.* (1) self-pollination or autogamy and (2) cross-pollination or allogamy.

6. When pollen grains are transferred from the anther to the stigma of same flower, it is known as self-pollination or autogamy.

7. Autogamy is the closest form of inbreeding. Autogamy leads to homozygosity.

8. Autogamy is promoted by mechanisms such as bisexuality, homogamy, cleistogamy, chasmogamy, *etc.*

9. Transfer of pollen grains from the anther of one plant to the stigma of another plant is called allogamy or cross pollination. This is the common form of outbreeding. Allogamy leads to heterozygosity.

10. Autogamy is promoted by mechanisms such as dicliny, dichogamy, heterostyly, herkogamy, self-incompatibility and male sterility.

11. When mate and female flowers are separate but present in the same plants, it is known as monoecy. In some crops, the male and female flowers are present in the same inflorescence such as in mango, castor and banana. In some cases, they are on separate inflorescence as in maize.

12. When staminate and pistillate flowers are present on different plants, it is called dioecy. In includes papaya, date palm, spinach, hemp and asparagus.

13. Dichogamy refers to maturation of anthers and stigma of the same flower at different times. Dichogamy promotes cross pollination even in the hermaphrodite species. Dichogamy is of two types: *viz.* (i) protogyny and (*ii*) protandry. When pistil matures before anthers, it is called protogyny such as in pearl millet. When anthers mature before pistil, it is known as protandry. It is found in maize, sugar beet and several other species.

14. When styles and filaments in a flower are of different length, it is called heterostyly. It promotes cross pollination, such as in Linseed.

15. Hindrance to self-pollination due to some physical barriers such as presence of hyline membrane around the anther is known as herkogamy. Such membrane does not allow the dehiscence of pollen and 'prevents self-pollinations such as in alfalfa.

16. The inability of fertile pollens to fertilize the same flower is referred to as self-incompatibility. It prevents self-pollination and promotes cross pollination. Self-incompatibility is found in several crop species like *Brassica,* Radish, *Nicotiana,* and many grass species. It is of two types sporophytic and gametophytic.

17. In some species, the pollen grains are non-functional. Such condition is known as male sterility. It prevents self-pollination and promotes cross pollination. It is of three types *viz.,* genetic, cytoplasmic and cytoplasmic genetic. It is a useful tool in hybrid seed production.

18. The mode of pollination plays an important role in plant breeding. It has impact on five aspects *viz.* (i) gene action, (ii) genetic constitution, (iii) adaptability, (iv) genetic purity and (v) transfer of genes.

19. On the basis of mode of pollination, crop plants can be classified into three groups: *viz.* (i) autogamous species, (ii) allogamous species and (iii) often allogamous species. In third group the cross pollination is found more than 5 per cent.

20. Based on mode of reproduction, crop plants can be divided into two groups: *viz.* (i) seed propagated and (ii) vegetatively propagated.

21. Asexual reproduction refers to multiplication of plants without the fusion of male and female gametes.

22. Crossing over does not take place in asexually reproducing species. In asexually propagating species variability is created only through somatic mutations.

23. Asexual reproduction is of two types: *viz.* (*a*) vegetative reproduction and (*b*) apomixis. Vegetative reproduction is again of two types: *viz.* (i) natural vegetative reproduction and (ii) artificial vegetative reproduction.

24. In nature, multiplication of certain plants occurs by underground stems, sub-aerial stems, roots and bulbils. In some crop species, underground stems (a modified group of stems) give rise to new plants.

25. Multiplication of plants by vegetative parts through artificial method is known as artificial vegetative reproduction. Such reproduction occurs by cuttings of stem and roots, and by layering, grafting and gootee. Examples of such reproduction are given below:
    Stem cuttings: Sugarcane (*Saccharum sp.*), grapes (*Vitis vinifera*), roses, *etc.*
    Root cuttings: Sweet potato, citrus, lemon, *etc.*

    Layering, grafting and gootee are used in fruit and ornamental crops.

26. Asexual reproduction leads to continuity of same genotype with great precision, because all the progeny of asexually reproduced plant will have similar genotype and phenotype.

27. Apomixis refers to the development of seed without sexual fusion (fertilization). In apomixis embryo develops without fertilization. Thus apomixis is an asexual means of reproduction. Apomixis is found in many crop species.

28. Reproduction in some species occurs only by apomixis. This apomixis is termed as obligate apomixis. But in some species sexual reproduction also occurs in addition to apomixis. Such apomixis is known as facultative apomixis.

29. There are four types of apomixis: *viz.* (1) parthenogenesis, (2) apogamy, (3) apospory and 4) adventive embryony.

30. There are several causes of parthenogenesis. The main causes includes: (i) inability of the pollen tube to discharge the contents inside the embryosac, (ii) insufficient attraction between male and female gametes, (iii) early degeneration of the sperm, (iv) very long style, (v) schlerenchymatous style. (vi) short pollen tube, (vii) slow rate of pollen tube growth, (viii) stimulation of pollination in the absence of pollen tube, and (ix) incompatibility.

31. Parthenogenesis can be artificially induced by four main ways : (i) by the stimulation of widely related pollen or foreign pollen, (ii) by low temperature, (iii) by pollinating with X-ray irradiated pollens, and (iv) by treatment with certain chemicals like belviton. All these help in inducing parthenogenetic development of egg cell.

32. The important applications include development of pure lines, conservation of heterosis and maintenance of superior genotypes.

C: MULTIPLE CHOICE QUESTIONS

1. Which of the following is often cross-pollinated crop?

(a) Wheat (b) Rice
(c) Cotton (d) Sesame

2. Which of the following is often cross-pollinated crop?

(a) Barley (b) Oats
(c) Pigeon pea (d) Chick pea

3. Which of the following is cross-pollinated crop?

(a) Lentil (b) Black gram
(c) Green gram (d) Maize

4. Which of the following is self-pollinated crop?

(a) Wheat (b) Rice
(c) Barley (d) All of these

5. Which of the following is cross-pollinated crop?

(a) Maize (b) Pearl millet
(c) Sunflower (d) All of these

6. Which of the following is cross-pollinated crop?

(a) Rye (b) Radish
(c) Alfalfa (d) All of these

7. Which of the following is not a cross-pollinated crop?

(a) Maize (b) Pearl millet
(c) Alfalfa (d) Peanut

8. Which of the following is not a cross-pollinated crop?

(a) Linseed (b) Sugar beet
(c) Carrot (d) Onion

9. Which of the following is not an often cross-pollinated crop?

(a) Wheat (b) Cotton
(c) Sorghum (d) Pigeon pea

10. Which of the following is often cross-pollinated crop?

(a) Triticale (b) Pigeon pea
(c) Tobacco (d) All of these

11. Potato propagates by:

(a) Bulbs (b) Rhizomes
(c) Corms (d) Tubers

12. Sweet potato propagates by:

(a) Stem cuttings (b) Rhizomes
(c) Corms (d) Root cuttings

13. Sugarcane propagates by:

(a) Stem cuttings (b) Rhizomes
(c) Corms (d) Root cuttings

14. Turmeric propagates by:

(a) Tubers (b) Rhizomes
(c) Corms (d) Bulbs

15. Arvi propagates by:

(a) Rhizomes (b) Bulbs
(c) Corms (d) Tubers

16. Garlic propagates by:

(a) Bulbs (b) Rhizomes
(c) Corms (d) Tubers

17. In which of the following both sexual and apomictic reproduction occur?

(a) Obligate apomixis (b) Facultative apomixis
(c) Recurrent apomixis (d) All of these

18. In parthenogenesis embryo develops from:

(a) Egg cell (b) Synergids
(c) Antipodal (d) None of these

19. In apogamy embryo develops from:

(a) Antipodal cell (b) Synergids
(c) Both (a) and (b) (d) Egg cell

20. In generative apospory embryo develops from:

(a) Egg cell (b) Synergids
(c) Antipodal (d) Archesporium

D: FILL IN THE BLANKS

1. Development of seed without fertilization is called________.
2. Development of seed with union of egg and sperm cell is called________.
3. Development of seed with self-pollination is known as________.
4. Development of seed with cross-pollination is known as________.
5. Development of embryo from egg cell refers to________.
6. Diploid apomixis is also called________.
7. Union of male and female gametes refers to________.
8. Maturation of anthers and stigma at the same time is called________.
9. Maturation of anthers and stigma at different time is called________.
10. Different lengths of styles and filaments in a flower refers to________.
11. The period of pollen distribution or dehiscence refers to________.
12. Sugarcane propagates by ________.
13. Potato propagates by ________.
14. Sweet potato propagates by ________.
15. Turmeric propagates by ________.
16. Ginger propagates by ________.
17. Garlic propagates by ________.
18. Arvi propagates by ________.
19. Pollination in unopened flower bud refers to ________.
20. Fertilization after opening of flower refers to ________.

E: MARK TRUE OR FALSE

1. Self-pollination is also known as autogamy.
2. Cross-pollination is also known as allogamy.
3. Self-pollination is a form of inbreeding.
4. Cross-pollination is a form of outbreeding.
5. Maturation of anthers and stigma at the same time is called homogamy.
6. Maturation of anthers and stigma at different time is called dichogamy.
7. Different lengths of styles and filaments in a flower refer to heterostyly.
8. Union of male and female gametes refers to amphimixis.
9. Apomixis does not involve union of male and female gametes.
10. In obligate apomixis reproduction occurs apomixis only
11. In facultative apomixis reproduction occurs both by sexual and apomictic means.
12. Autogamy leads to homozygosity.
13. Allogamy leads to heterozygosity.
14. In cleistogamy, fertilization occurs in unopened flower bud.
15. In chasmogamy, fertilization occurs after opening of flower bud.
16. Presence of a hyline membrane around anther refers to herkogamy.
17. Sugarcane propagates by stem cuttings.
18. Potato propagates by tubers.
19. Sweet potato propagates by root cuttings.
20. Turmeric propagates by Rhizomes.

F: ONE WORD ANSWER

1. The process by which living organisms give rise to the offspring of similar kind (species).
2. Multiplication of plants through fertilized embryos.
3. Development of seed by self-pollination.
4. Development of seed by cross pollination.

5. Presence of male and female organs in the same flower.
6. Maturation of anthers and stigma of a flower at the same time.
7. Completion of pollination and fertilization in unopened flower bud.
8. Fertilization after opening of flower.
9. Separate existence of male and female flowers on the same plant in the same inflorescence or different inflorescence.
10. Existence of male and female flowers on different plants.
11. Maturation of anthers and stigma of a flower at different time.
12. Different lengths of styles and filaments in a flower.
13. Hindrance to self-pollination due to presence of hyline membrane around anther.
14. The process of dehiscence of anthers, the period of pollen distribution.
15. Multiplication of plants without the fusion of male and female gametes. It is of two type, *viz.*: vegetative reproduction and apomixis.
16. Multiplication of plants by means of various vegetative plant parts such as stem cuttings and root cuttings.
17. Vegetative reproduction which occurs in nature.
18. Vegetative reproduction which is achieved by artificial means.
19. Development of seed without sexual fusion (fertilization).
20. Apomixis in which reproduction occurs only by apomixis.
21. Apomixis in which sexual reproduction occurs in addition to apomixis.
22. Occurrence of apomixis from one generation to another.
23. When apomictic plants are not obtained from one generation to another.
24. Development of embryo from the egg cell without fertilization.
25. Development of embryo from haploid egg cell.
26. Development of embryo from diploid egg cell.
27. Development of embryo either from synergids or antipodal cells of embryosac.

28. Development of another embryosac without reduction from the cell of ovule outside the embryosac and then development of embryo directly from diploid egg cell.

29. Development of new embryosac from archesporium.

30. Development of new embryosac from nucellus or integument.

31. Development of embryo from the diploid cells of ovule lying outside the embryosac belonging to either nucellus or integument.

ANSWERS

MULTIPLE CHOICE QUESTIONS

Sl. No.	*Answer*	*Sl. No.*	*Answer*
1	(c) Cotton	2	(c) Pigeon pea
3	(d) Maize	4	(d) All of these
5	(d) All of these	6	(d) All of these
7	(d) Peanut	8	(a) Linseed
9	(a) Wheat	10	(d) All of these
11	(d) Tubers	12	(d) Root cuttings
13	(a) Stem cuttings	14	(b) Rhizomes
15	(b) Rhizomes	16	(a) Bulbs
17	(b) Facultative apomixis	18	(a) Egg cell
19	(c) Both (a) and (b)	20	(d) Archesporium.

FILL IN THE BLANKS

Sl. No.	*Answer*	*Sl. No.*	*Answer*
1	Apomixis	2	Sexual Reproduction
3	Autogamy	4	Allogamy
5	Parthenogenesis	6	Recurrent apomixis
7	Amphimixis	8	Homogamy
9	Dichogamy	10	Hetrostyly
11	Anthesis	12	Stem cuttings
13	Tubers	14	Root cuttings
15	Rhizomes	16	Rhizomes
17	Bulbs	18	Corms
19	Cleistogamy	20	Chasmogamy

TRUE OR FALSE

1 to 20 all are true.

APPROPRIATE TERM/ONE WORD ANSWER

Sl. No.	*Answer*	*Sl. No.*	*Answer*
1	Reproduction	2	Sexual Reproduction
3	Autogamy	4	Allogamy
5	Bisexuality	6	Homogamy
7	Cleistogamy	8	Chasmogamy
9	Monoecy	10	Dioecy
11	Dichogamy	12	Heterostyly
13	Herkogamy	14	Anthesis
15	Asexual Reproduction	16	Vegetative Reproduction
17	Natural Vegetative Reproduction	18	Artificial Vegetative Reproduction
19	Apomixis	20	Obligate Apomixix
21	Facultative Apomixis	22	Recurrent Apomixis
23	Non-Recurrent Apomixis	24	Parthenogenesis
25	Haploid Parthenogenesis	26	Diploid Parthenogenesis
27	Apogamy	28	Apospory
29	Generative Apospory	30	Somatic Apospory
31	Adventive Embryony		

*****

Chapter 2

Types of Improved Seed

A: DEFINITIONS

Breeder seed: The progeny of either nucleus seed or breeder seed produced under the direct supervision of original or sponsoring plant breeder.

Certified-seed: The progeny of either foundation seed or certified seed.

Foundation seed: The progeny of breeder seed produced by NSC or SSC.

Improved Seed: The seed of a released and popular variety produced by scientific method; also called quality seed.

Nucleus seed: The initial seed of a released and notified variety which is always limited in quantity.

Registered- seed: The progeny of either foundation seed or registered seed. This category is omitted in India.

Variety: A genotype which has been released for commercial cultivation either by State Variety Release Committee or Central Variety Release Committee.

B: BACKGROUND INFORMATION

1. A new variety is released either by State Variety Release Committee or Central Variety Release Committee.

2. Improved seed plays an important role in maximizing production and productivity of field crops. It results in (i) better germination, (ii) vigorous seedling growth, (iii) higher crop stand, (iv) better quality of produce, and (v) ultimately in higher crop yield.

3. Seed quality is measured in terms of genetic purity, physical purity, germination and seed health.

4. The seeds that meet required standard of genetic purity, good health and physiological purity (viability and vigor) and other attributes are referred to as quality seed.

5. There are certain characteristics of quality seeds such as released variety, genetic purity, seed health, free from inert matter and mixture, undamaged seeds, seed viability, *etc.*

6. There are five classes or categories of improved seed, *viz.*, nucleus seed, breeder seed, foundation seed, registered seed and certified seed.

7. Certification is not required for the production of nucleus seed and breeder seed.

8. Nucleus seed is also known as basic seed or primordial seed.

9. Nucleus seed is produced at the research farms of originating research institute or agricultural university.

10. The color of seed bags of breeder seed is yellow.

11. The color of seed bags of foundation seed is white.

12. In India, registered seed is omitted and certified seed is produced directly from foundation seed.

13. The color of seed bags of certified seed is blue.

14. The certified seed is of two types, *viz.*, certified seed stage I and stage II. The former is produced from foundation seed and latter from certified seed stage I.

15. Certified seed is available to the farmers for commercial crop production.

C. MULTIPLE CHOICE QUESTIONS

1. The initial seed lot of a variety is called:

(a) Nucleus seed (b) Breeder seed
(c) Foundation seed (d) Registered seed

2. Nucleus seed is also known as:

(a) Basic seed (b) Primordial seed
(c) Both (a) and (b) (d) None of these

3. Breeder seed is the progeny of:

(a) Nucleus seed (b) Breeder seed
(c) Foundation seed (d) Registered seed

4. Foundation seed is the progeny of:

(a) Nucleus seed (b) Breeder seed

(c) Foundation seed (d) Registered seed

5. Certified seed is the progeny of:

(a) Nucleus seed (b) Breeder seed

(c) Foundation seed (d) Registered seed

6. Which category of seed is omitted in India?

(a) Nucleus seed (b) Breeder seed

(c) Foundation seed (d) Registered seed

7. Which category of seed is to provided to farmers for commercial crop production?

(a) Nucleus seed (b) Breeder seed

(c) Both a and b (d) Registered seed

9. Color of breeder seed tag is:

(a) Yellow (b) White

(c) Blue (d) Green

10. Color of foundation seed tag is:

(a) Yellow (b) White

(c) Blue (d) Green

11. Color of certified seed tag is:

(a) Yellow (b) White

(c) Blue (d) Green

D. FILL IN THE BLANKS

1. The initial seed lot of a variety is called ________seed.
2. Nucleus seed is also known as basic seed or ________ seed.
3. Breeder seed is the progeny of ________seed.
4. Foundation seed is the progeny of ________seed.
5. Certified seed is the progeny of ________ seed.
6. Nucleus and breeder seeds are produced by ________ breeder.
7. Foundation seed is produced by ________.
8. Certified seed is produced by ________.

9. ________ seed category is omitted in India.
10. Certification is not required in ________and ________seeds.
11. Color of the tag of breeder seed is ________.
12. Color of foundation seed is ________.
13. Color of certified seed is ________.
14. Certified seed stage I is produced from ________seed.
15. Certified seed stage II is produced from certified seed ________.
16. ________seed is available to the farmers for commercial crop production.

E. MARK TRUE OR FALSE

1. The initial seed lot of a variety is called nucleus seed.
2. Nucleus seed is also known as basic seed or primordial seed.
3. Breeder seed is the progeny of nucleus seed.
4. Foundation seed is the progeny of breeder seed.
5. Certified seed is the progeny of foundation seed.
6. Nucleus and breeder seeds are produced by original or sponsored breeder.
7. Foundation seed is produced by NSC.
8. Certified seed is produced by State Seeds Corporations.
9. Registered seed category is omitted in India.
10. Certification is not required in nucleus and breeder seeds.
11. Color of the tag of breeder seed is yellow.
12. Color of foundation seed is white.
13. Color of certified seed is blue.
14. Certified seed stage I is produced from foundation seed.
15. Certified seed stage II is produced from certified seed stage I.
16. Certified seed is available to the farmers for commercial crop production.

F. ONE WORD ANSWER

1. The seed of a release and popular variety produced by scientific method; also called quality seed.
2. A genotype which has been released for commercial cultivation either by State Variety Release Committee or Central Variety Release Committee.
3. The initial seed of a released and notified variety which is always limited in quantity.
4. The progeny of either nucleus seed or breeder seed produced under the direct supervision of original or sponsoring plant breeder.
5. The progeny of breeder seed produced by NSC or SSC.
6. The progeny of either foundation seed or registered seed. This category is omitted in India.
7. The progeny of either foundation seed or certified seed.

ANSWERS

MULTIPLE CHOICE QUESTIONS

Sl. No.	*Answer*	*Sl. No.*	*Answer*
1	(a) Nucleus seed	7	(d) Certified seed
2	(c) Both a and b	8	(c) Both a and b
3	(a) Nucleus seed	9	(a) Yellow
4	b) Breeder seed	10	(b) White
5	(c) Foundation seed	11	(c) Blue
6	(d) Registered seed		

FILL IN THE BLANKS

Sl. No.	*Answer*	*Sl. No.*	*Answer*
1	Nucleus	9	Registered
2	Primordial	10	Nucleus, breeder
3	Nucleus	11	Yellow
4	Breeder	12	White
5	Foundation	13	Blue
6	Original/sponsored	14	Foundation
7	NSC	15	Certified stage I
8	SSCs	16	Certified

TRUE OR FALSE

1 to 20 all are true.

APPROPRIIATE TERM/ONE WORD ANSWER

Sl. No.	*Answer*	*Sl. No.*	*Answer*
1	Improved Seed	5	Foundation seed
2	Variety	6	Registered seed
3	Nucleus seed	7	Certified seed
4	Breeder seed		

*****

Chapter 3

Development of Inbred Lines

A: DEFINITIONS

Double Cross Hybrid: The hybrid progeny from a cross between two single crosses. *i.e.* (A x B) x (C x D).

Double Top-cross: A cross between a single cross and an open pollinated variety, *viz.* (A x B) x OP variety.

Hybrid: The progeny of a cross between genetically dissimilar parents.

Hybrid Variety: The F_1 population that is used for commercial cultivation.

Inbred: In cross pollinated species, a true breeding line that is obtained by continuous inbreeding

Intraspecific Cross: A cross between two varieties or genotypes of the same species; also called inter-varietal cross.

Interspecific Cross: A cross between two different species of the same genus.

Multiple Cross: A cross involving more than four inbred lines; also known as composite cross.

Poly cross: Open pollination of a group of selected genotypes in isolation from other compatible genotypes to promote random mating among selected genotypes.

Single Cross Hybrid: The hybrid progeny from a cross between two inbreds or varieties, *viz.* (A x B).

Three way Cross Hybrid: The hybrid progeny between a single cross and an inbred *viz.* (A x B) x C.

Top Cross: A cross between an inbred line and an open pollinated variety; also known as inbred variety cross.

B: BACKGROUND INFORMATION

1. Progeny of a self-pollinated species obtained by selfing is called pure line.
2. All the genotypes of a pure line are genetically and phenotypically similar.
3. Pure lines are isolated from heterogeneous population and segregating population.
4. A pure line has narrow adaptation due to poor buffering capacity.
5. Natural out-crossing, mechanical mixture and mutations are sources of variation in pure line variety.
6. In cross pollinated species, a true breeding line obtained by continuous inbreeding is called inbred line. Development of inbred lines is an important step in the production of hybrids.
7. Inbred lines are used in hybridization programs for development of single, three-way and double cross hybrids.
8. There are two methods of developing inbred lines, *viz.* by selfing of heterozygous populations and by doubling of haploids.
9. Two methods, *viz.* (i) top cross method, and (ii) single crosses are commonly used for evaluation of inbred lines in terms of their combining ability or performance in hybrid combinations.
10. The testing of inbred lines for general combining ability should be started from 3rd, 4th and 5th generation of selfing. This will help in retaining of inbred lines with good combining ability and elimination of lines with poor combining ability.
11. Inbred lines are used in developing productive open pollinated cultivars, synthetic and composite cultivars and single, three-way and double cross hybrids.
12. Single cross hybrid involves two different inbred or pure lines.
13. Three-way cross hybrid involves three different genotypes.
14. Double cross hybrid involves four different genotypes.
15. Composite cross hybrid involves more than four different genotypes.
16. Composite cross is also known as multiple cross.
17. Inbreed lines are related to cross pollinated crops.
18. Pure lines are related to self-pollinated crops.

19. Inbred lines are developed by inbreeding.
20. Pure lines are developed by selfing in self-pollinated species.

C: MULTIPLE CHOICE QUESTIONS

1. Inbreeding tends to increase in:

(a) Homozygosity (b) Genetic correlation
(c) Genetic diversity (d) All of these

2. Out-breeding tends to reduce in:

(a) Homozygosity (b) Genetic correlation
(c) Genetic variability (d) All of these

3. Inbreeding leads to reduce in:

(a) General vigor (b) Adaptation
(c) Genetic diversity (d) All of these

4. A cross between two inbred lines is called:

(a) Single cross (b) Three-way cross
(c) Double cross (d) Multiple-cross

5. In which cross highest uniformity is observed?

(a) Single cross (b) Three-way cross
(c) Double cross (d) Multiple-cross

6. An inbred-variety cross refers to:

(a) Three-way cross (b) Double cross
(c) Top cross (d) Poly cross

7. A cross between single cross and open pollinated variety refers:

(a) Poly cross (b) Composite cross
(c) Top cross (d) Double top cross

8. Hybrid vigor results due to:

(a) Dominance (b) Over-dominance
(c) Epistasis (d) All of these

9. In-breeders have:

(a) Selfing (b) Homozygosity
(c) Close adaptation (d) All of these

10. Out-breeders have:

(a) Random mating (b) Heterozygosity

(c) Close adaptation (d) All of these

11. Hybrid vigor can be conserved by:

(a) Asexual reproduction (b) Apomixis

(c) Polyploidy (d) All of these

D: FILL IN THE BLANKS

1. Inbred lines are developed by ________.
2. Pure lines are developed by ________.
3. Inbred lines are related to ________species.
4. Pure lines are related to ________species.
5. Self-pollinated species are also known as ________.
6. Cross-pollinated species are also called ________.
7. Single cross hybrid involves two different ________ lines.
8. Double cross hybrid is a cross between ________F_1 crosses.
9. Composite cross hybrid involves more than ________ genotypes.
10. Composite cross is also known as ________ cross.
11. A three-way cross involves ________genotypes.
12. A double cross hybrid involves ________genotypes.
13. All the genotypes of a pure line are genetically and phenotypically ________.
14. A pure line has narrow adaptation due to poor ________ capacity.

E: MARK TRUE OR FALSE

1. Inbred lines are developed by inbreeding.
2. Pure lines are developed by selfing in self-pollinated species.
3. Inbred lines are related to outbreeders.
4. Pure lines are related to inbreeders.

5. Self-pollinated species are also known as inbreeders.
6. Cross-pollinated species are also called outbreeders.
7. Single cross hybrid involves two different inbred or pure lines.
8. Three-way cross hybrid involves three different genotypes.
9. Double cross hybrid involves four different genotypes.
10. Composite cross hybrid involves more than four different genotypes.
11. Composite cross is also known as multiple cross.
12. Inbreed lines are related to cross pollinated crops.
13. Pure lines are related to self-pollinated crops.
14. All the genotypes of a pure line are genetically and phenotypically similar.
15. A pure line has narrow adaptation due to poor buffering capacity.

F: ONE WORD ANSWER

1. The progeny of a cross between genetically dissimilar parents.
2. The F_1 population that is used for commercial cultivation.
3. The hybrid progeny from a cross between two inbreds or varieties, *viz.* (A x B).
4. The hybrid progeny between a single cross and an inbred *viz.* (A x B) x C.
5. The hybrid progeny from a cross between two single crosses., *i.e.* (A x B) x (C x D).
6. A cross between an inbred line and an open pollinated variety; also known as inbred variety cross.
7. A cross between a single cross and an open pollinated variety, *viz.* (A x B) x OP variety.
8. A cross involving more than four inbred lines; also known as composite cross.
9. Open pollination of a group of selected genotypes in isolation from other compatible genotypes to promote random mating among selected genotypes.

10. A cross between two varieties or genotypes of the same species; also called inter-varietal cross.

11. A cross between two different species of the same genus.

12. In cross pollinated species, a true breeding line that is obtained by continuous inbreeding

ANSWERS

MULTIPLE CHOICE QUESTIONS

Sl. No.	*Answer*	*Sl. No.*	*Answer*
1	(d) All of these	2	(d) All of these
3	(d) All of these	4	(a) Single cross
5	(a) Single cross	6	(c) Top cross
7	(d) Double top cross	8	(d) All of these
9	(d) All of these	10	(d) All of these
11	(d) All of these		

FILL IN THE BLANKS

Sl. No.	*Answer*	*Sl. No.*	*Answer*
1	Inbreeding	2	Selfing
3	Cross-pollinated	4	Self-pollinated
5	Inbreeders	6	Outbreeders
7	Inbred or pure lines	8	Two
9	Four	10	Multiple cross
11	Three	12	Four
13	Similar	14	Buffering

TRUE OR FALSE

1 to 15 all are true.

APPROPRIATE TERM/ONE WORD ANSWER

Sl. No.	*Answer*	*Sl. No.*	*Answer*
1	Hybrid	2	Hybrid variety
3	Single Cross Hybrid	4	Three way Cross Hybrid
5	Double Cross Hybrid	6	Top Cross
7	Double Top-cross	8	Multiple Cross
9	Poly cross	10	Intraspecific Cross
11	Interspecific Cross	12	Inbred

*****

Chapter 4

Development of Hybrids using A, B and R Lines

A: DEFINITIONS

Double Cross Hybrid: The hybrid progeny from a cross between two single crosses., *i.e.* (A x B) x (C x D).

Double Top-cross: A cross between a single cross and an open pollinated variety, *viz.* (A x B) x OP variety.

Hybrid Variety: The F_1 population that is used for commercial cultivation.

Hybrid: The progeny of a cross between genetically dissimilar parents.

Inbred: In cross pollinated species, a true breeding line obtained by continuous inbreeding.

Interspecific Cross: A cross between two different species of the same genus.

Intraspecific Cross: A cross between two varieties or genotypes of the same species; also called inter-varietal cross.

Multiple Cross: A cross involving more than four inbred lines; also known as composite cross.

Poly cross: Open pollination of a group of selected genotypes in isolation from other compatible genotypes to promote random mating among selected genotypes.

Single Cross Hybrid: The hybrid progeny from a cross between two inbreds or varieties, *viz.* (A x B).

Three way Cross Hybrid: The hybrid progeny between a single cross and an inbred *viz.* (A x B) x C.

Top Cross: A cross between an inbred line and an open pollinated variety; also known as inbred variety cross.

B: BACKGROUND INFORMATION

1. In practical plant breeding, heterosis can be fully exploited in the form of hybrids, and partially in the form of synthetic and composite varieties.

2. The commercial exploitation of heterosis depends on magnitude of heterosis, outcrossing percent, floral biology and availability of male sterility system.

3. The level of heterosis differs from crop to crop. It is generally higher in cross pollinated species than in self-pollinated species.

4. Heterosis of 40 per cent and above is considered as significant in most of the crops.

5. Availability of male sterility (*MS*) and self-incompatibility (*SI*) systems help in reducing the cost of hybrid seed by eliminating the process of hand emasculation.

6. There should be proper nicking in the flowering time of A and *R* lines.

7. The *A* line should have good stigma receptivity and *R* line should have high pollen production efficiency and pollen longevity.

8. In case of self-incompatibility, the modifier genes which break self-incompatibility should be absent.

9. The commercially cultivated hybrids are of two types, *viz.* (a) intra-specific hybrids, and (b) inter-specific hybrids.

10. A hybrid between genetically different genotypes of the same species is known as intra-specific hybrid. It is also known as inter-varietal hybrid.

11. Intra-specific hybrids are always fertile. Based on type of cross, intra-specific hybrids are of three types, *viz.* (a) single cross hybrid, (*b*) three-way cross hybrid, and (c) double cross hybrid.

12. The F_1 progeny between two different species of the same genus is known as interspecific hybrid. This is also referred to as intrageneric hybrid.

13. Interspecific hybrids are rarely used for commercial cultivation, because such hybrids are fertile only in few cases.

13. In cotton, interspecific hybrids between tetraploid cultivated species (*Gossypium hirsutum* x *G. barbadense*) and diploid cultivated species (*G. arboreum* x *G. herbaceum*) are fully fertile. Several interspecific hybrids at tetraploid level (Varalaxmi, DCH 32, HB 224, NHB12, TCHB 213) and some hybrids at diploid level (DH7, DH9, and DDH2) have been released for commercial cultivation in India, besides large number of intraspecific hirsutum hybrids.

14. In cotton, male sterility based hybrids include PKV Hy3, PKV Hy4. PKV Hy5 at tetraploid level. All these are CMS based. Diploid hybrids AKDH7, AAH1, G. Cot. MDH 11 and CICR hy2 are GMS based.

15. In maize, the production of hybrid varieties in maize consists of three main steps, *viz.* (1) development of inbred lines, (2) selection of productive inbred lines, and (3) production of hybrid seed.

16. The use of cytoplasmic male sterility helps in reducing the cost of hybrid seed by eliminating the process of detasseling. In maize, the CMS can be used in three different ways in the production of double cross hybrids.

17. In cross pollinated crops like maize and pearl millet, heterosis is utilized fully as well as partially. But in often cross pollinated species and self-pollinated crops heterosis is fully exploited by developing high yielding hybrids.

18. Important maize hybrids include Ganga 2, Ganga 5, Ganga 11, Daccan 103, Makka 1 and VL 42.

19. Important pearl millet hybrids include HB 3, BJ 104, MHB 110, BK 560 and MH 179.

20. Important sunflower hybrids include BSH 1, HSFH 1, MSFH 8, MSFH 10, KBSH 1 and KBSH 11.

21. Important *Sorghum* hybrids include CSH 1, CSH 5, CSH 6, CSH 9, CSH 10 and CSH 11.

22. Important cotton hybrids include H 4, H 6, H8, H10, JKHY 1, NHH 44, Savita, Fateh, Raj HH 16, Dhanlaxmi, Varalaxmi, DCH 32, HB 224, TCHB 213, DH 7, DH 9, DDH 2, Bt. Hybrids, *etc.*

23. In pigeon pea, ICPH 8 was the first hybrid which was developed from ICRISAT, Hyderabad in 1991 using GMS line.

24. In Rice first hybrid was developed in China by Yuan Long Ping *et al.*, in 1976 using CMS line.

25. In cotton, Worlds first hybrid was developed in 1970 by Dr. C. T. Patel from main Cotton Research Station, Surat, Gujarat.

26. Hybrid varieties have higher yield potential as compared to synthetics, composites and open pollinated varieties.

27. Hybrid varieties are more uniform and attractive than synthetics, composites and open pollinated varieties.

C: MULTIPLE CHOICE QUESTIONS

1. The World's first cotton hybrid is:

(a) Hybrid 4 (b) Hybrid 6

(c) Hybrid 8 (d) Hybrid 10

2. The first GMS based cotton hybrid is:

(a) Hybrid 6 (b) Suguna

(c) Hybrid 8 (d) Hybrid 10

3. Which of the following cotton hybrids is CMS based?

(a) PKV Hybrid 3 (b) PKV Hybrid 4

(c) PKV Hybrid 5 (d) All of these

4. Which of the following cotton hybrids is not CMS based?

(a) PKV Hybrid 3 (b) PKV Hybrid 4

(c) PKV Hybrid 5 (d) Hybrid 4

5. In which year hybrid 4 was released for commercial cultivation?

(a) 1968 (b) 1970

(c) 1972 (d) 1975

6. Where was first rice hybrid developed?

(a) India (b) Philippines

(c) China (d) Japan

7. Which of the following is first pigeon pea hybrid?

(a) CPH 3 (b) CPH 5

(c) CPH 8 (d) None of these

8. Which of the following States is home of hybrid cotton?

(a) Tamil Nadu (b) Gujarat

(c) Punjab (d) None of these

9. Which of the following is *Sorghum* hybrid?

(a) H4 (b) CPH 8

(c) CSH 5 (d) None of these

10. Who is known as father of hybrid cotton?

(a) V. Santhanam (b) C. T. Patel

(c) N. P. Mehta (d) None of these

D: FILL IN THE BLANKS

1. In cotton, ________ was the first hybrid released for commercial cultivation.
2. Cotton hybrid H4 was developed by Dr. ________.
3. ________ is the World's first cotton hybrid.
4. H4 was released for commercial cultivation in ________
5. Dr. ________ is known as father of hybrid cotton.
6. In cotton, ________was the first GMS based hybrid.
7. ________State is known as home of the hybrid cotton.
8. Cotton hybrid PKV Hy3 is ________ based.
9. Cotton hybrid PKV Hy4 is ________ based.
10. Cotton hybrid PKV Hy5 is ________ based.
11. Cotton hybrid AKDH7 is ________ based.
12. Cotton hybrid AAH1 is ________ based.
13. Cotton hybrid G. Cot. MDH 11 is ________ based.
14. Cotton hybrid CICR hy2 is ________ based.
15. In rice, first CMS based hybrid was developed in ________.
16. In Rice, first hybrid was developed in China by ________.
17. The first Pigeon pea hybrid is ________.
18. Pigeon pea hybrid CPH 8 was developed from ________.
19. Pigeon pea hybrid CPH8 is ________ based.
20. CSH 5 is a ________hybrid.

E: MARK TRUE OR FALSE

1. In cotton, H4 was the first hybrid released for commercial cultivation.
2. Cotton hybrid H4 was developed by Dr. C. T. Patel.
3. H4 is the World's first cotton hybrid.
4. H4 was released for commercial cultivation i9 in 1970
5. Dr. C. T. Patel is known as father of hybrid cotton.

6. In cotton, Suguna was the first GMS based hybrid.
7. Gujarat State is known as home of the hybrid cotton.
8. Cotton hybrid PKV Hy3 is CMS based.
9. Cotton hybrid PKV Hy4 is CMS based.
10. Cotton hybrid PKV Hy5 is CMS based.
11. Cotton hybrid AKDH7 is GMS based.
12. Cotton hybrid AAH1 is GMS based.
13. Cotton hybrid G. Cot. MDH 11 is GMS based.
14. Cotton hybrid CICR hy2 is GMS based.
15. In rice, first CMS based hybrid was developed in China.
16. In Rice, first hybrid was developed in China by Yuan Long Ping *et al.*
17. The first Pigeon pea hybrid is CPH8.
18. Pigeon pea hybrid CPH 8 was developed from ICRISAT.
19. Pigeon pea hybrid CPH8 is GMS based.
20. CSH 5 is a Sorghum hybrid.

F: ONE WORD ANSWER

1. The progeny of a cross between genetically dissimilar parents.
2. The F_1 population that is used for commercial cultivation.
3. The hybrid progeny from a cross between two inbreds or varieties, *viz.* (A x B).
4. The hybrid progeny between a single cross and an inbred, *viz.* (A x B) x C.
5. The hybrid progeny from a cross between two single crosses, *i.e.* (A x B) x (C x D).
6. A cross between an inbred line and an open pollinated variety; also known as inbred variety cross.
7. A cross between a single cross and an open pollinated variety, *viz.* (A x B) x OP variety.
8. A cross involving more than four inbred lines; also known as composite cross.

9. Open pollination of a group of selected genotypes in isolation from other compatible genotypes to promote random mating among selected genotypes.

10. A cross between two varieties or genotypes of the same species; also called inter-varietal cross.

11. A cross between two different species of the same genus.

12. In cross pollinated species, a true breeding line obtained by continuous inbreeding

ANSWERS

MULTIPLE CHOICE QUESTIONS

Sl. No.	*Answer*	*Sl. No.*	*Answer*
1	(a) Hybrid 4	2	(b) Suguna
3	(d) All of these	4	(d) Hybrid 4
5	(b) 1970	6	(c) China
7	(c) CPH 8	8	(b) Gujarat
9	(c) CSH 5	10	(b) C. T. Patel

FILL IN THE BLANKS

Sl. No.	*Answer*	*Sl. No.*	*Answer*
1	H4	2	C. T. Patel
3	H4	4	1970
5	C. T. Patel	6	Suguna
7	Gujarat	8	CMS
9	CMS	10	CMS
11	GMS	12	GMS
13	GMS	14	GMS
15	China	16	Yuan Long Ping *et al.*
17	CPH 8	18	ICRISAT
19	GMS	20	*Sorghum*

TRUE OR FALSE

1 to 20 all are true.

APPROPRIATE TERM/ONE WORD ANSWER

Sl. No.	*Answer*	*Sl. No.*	*Answer*
1	Hybrid	2	Hybrid Variety
3	Single Cross Hybrid	4	Three way Cross Hybrid
5	Double Cross Hybrid	6	Top Cross
7	Double Top-cross	8	Multiple Cross
9	Poly cross	10	Intraspecific Cross
11	Interspecific Cross	12	Inbred

*****

Chapter 5

Variety Release Procedure

A: DEFINITIONS

Clonal variety: A variety which is developed by clonal selection, relevant to asexually propagated species.

Hybrid variety: A F_1 population which is used for commercial cultivation.

Mutant variety: A variety which is developed by mutation breeding.

Multi-line variety: A variety of self-pollinated crops which is developed by mixing seed of several isogenic lines, related lines or unrelated lines in equal quantity.

Pure line variety: A variety which is developed by pure line selection.

Pedigree bred variety: A variety which developed by pedigree breeding method.

Strain: A group of genotypes that are similar in phenotype. After release a strain is called variety.

Transgenic variety: A variety which is developed by the technique of genetic engineering. It differs in one character from the parent variety.

Variety: A strain which is released for commercial cultivation by a variety release committee.

B: BACKGROUND INFORMATION

1. A new variety is released for commercial cultivation either by State Variety Release Committee (SVRC) or Central Variety Release Committee (CVRC) and notified by the government of India.
2. A variety is also termed as cultivar (cultivated variety).
3. A variety has three important characteristics, *viz.* (1) distinctness, (2) uniformity, and (3) stability.

4. The distinctness means that the variety should be distinguishable by one or more morphological, physiological and biochemical features from all pre-existing cultivars.

5. Uniformity means that the variety should be pure and look uniform.

6. Stability means that the variety should give stable performance in different generations. In other words, the variety should remain unchanged in its distinctness and uniformity for reasonable period of time when reproduced or reconstituted (hybrid).

7. The new variety should be superior to the previously released varieties of a crop in yield potential of grain, oil, fibre, fodder, vegetable or other economic product. It should have resistance to biotic (diseases, insects and parasitic weeds) and abiotic (drought, salinity, frost, cold, heat, metal toxicity, *etc.*) stresses.

8. It may be superior in quality of oil, fibre, protein, fodder, vegetable, grain, *etc.*, maturity duration, adaptability or suitability for machine harvesting, *etc.*

9. In plant breeding various types of varieties such as pure line variety, mass selected variety, clonal variety, synthetic variety, composite variety, mutant variety, transgenic variety and hybrid variety are used.

10. In India, new crop cultivars are developed by ICAR crop Research Institutes, SAUs and Private Seed Companies.

11. In India, the release of new crop varieties consists of four major steps, *viz.* (1) development of new strains, (2) evaluation of performance, (3) identification of superior strains, and (4) release and notification.

12. The new strains are developed by ICAR Crop Research Institutes and State Agricultural Universities for specific purposes. Various breeding methods are used for development of new strains in self and cross pollinated species.

13. The performance of newly developed strains is evaluated in AICCIP. ICAR Institutes, SAUs and private registered seed companies enter their improved strains/hybrids in the AICCIP of respective crop for multi-location testing. The new strains are tested at multi-locations under the coordinated project for a minimum period of three years/seasons.

14. The new variety is first tested for yield under the initial varietal trials (IVT); the promising ones are promoted to the advanced varietal trials (AVT) in the second year. The strains that give good performance in AVT for two years are selected.

15. The strains which show good yield performance in AVT are identified as superior strains and are considered for release in the workshop meetings.

16. The newer agro and plant protection techniques required to obtain potential yield of new strains are also worked out by that time.

17. The workshop after considering the new promising varieties recommends them to replace existing varieties.

18. The proposal for release of new varieties is put up in a prescribed format to variety release committee. There are two types of variety release committees, *viz.* State Variety Release Committee (SVRC) and Central Variety Release Committee (CVRC).

19. In case of state variety release committee, Director of Agriculture for field crops and Director of horticulture for vegetable and horticulture crops is the chairman.

20. In central variety release committee, Deputy Director General (Crop Science) of ICAR is the Chairman. The release proposal of varieties recommended for All India release is put up before CVRC, while for those recommended for release in a particular state is placed before the SVRC of respective state.

21. These committees consist of scientists, and representatives of seed producing organizations (NSC, SSC and SSCA) and other related government agencies. After release, the variety is notified.

22. Seed production can be taken up only after notification of new varieties. The notification is done by the government of India.

C: MULTIPLE CHOICE QUESTIONS

1. Who is the chairman of Central Variety Release Committee?

(a) Director of Agriculture (b) DDG Crop Science, ICAR

(c) Secretary ICAR (d) None of these

2. Who is the chairman of State Variety Release Committee for field crops?

(a) Director of Agriculture (b) DDG Crop Science, ICAR

(c) Secretary ICAR (d) None of these

3. New crop cultivars can be developed by:

(a) ICAR Crop Res. Institutes (b) SAUs

(c) Private Seed Companies (d) All of these

4. For release, the new variety should be superior to existing ones in:

(a) Yield potential
(b) Quality of produce
(c) Resistance to insects and disease
(d) Any one of these

5. Seed production of the new variety can be taken up:

(a) After release
(b) After registration
(c) Before registration
(d) All of these

6. Cultivar may include:

(a) Pure line variety
(b) Mass selected variety
(c) Clonal variety
(d) All of these

7. Cultivar may include:

(a) Hybrid variety
(b) Composite
(c) Synthetic variety
(d) All of these

8. Which of the following is developed using biotechnology?

(a) Hybrid variety
(b) Mutant variety
(c) Clonal variety
(d) Transgenic variety

9. Golden Rice has been developed by using which method?

(a) Pure line selection
(b) Mass selection
(c) Recurrent selection
(d) Transgenic breeding

10. Which of the following methods involves protoplast fusion?

(a) Pure line selection
(b) Varietal hybridization
(c) Somatic hybridization
(d) All of these

D: FILL IN THE BLANKS

1. A variety is also known as ________.
2. New variety is notified by ________.
3. Seed production of new variety can be taken up only after ________.
4. In central variety release committee, ________ of ICAR is the Chairman.
5. The release proposal of varieties recommended for All India release is put up before ________.
6. The release proposal of varieties recommended for release in a particular state is placed before the ________ of respective state.
7. In case of state variety release committee, ________ is chairman for field crops.

8. In case of state variety release committee ________ is chairman for vegetable and horticulture crops.

9. The proposal for release of new varieties is put up in a prescribed format to ________.

10. The strains which show good yield performance in ________ are identified as superior strains and are considered for release in the workshop meetings.

E: MARK TRUE OR FALSE

1. A variety is also known as cultivar.
2. New variety is notified by government of India.
3. Seed production of new variety can be taken up only notification.
4. In central variety release committee, Deputy Director General (Crop Science) of ICAR is the Chairman.
5. The release proposal of varieties recommended for All India release is put up before CVRC.
6. The release proposal of varieties recommended for release in a particular state is placed before the SVRC of respective state.
7. In case of state variety release committee, Director of Agriculture is chairman for field crops.
8. In SVRC Director of horticulture is chairman for vegetable and horticulture crops.
9. The proposal for release of new varieties is put up in a prescribed format to variety release committee.
10. The strains which show good yield performance in AVT are identified as superior strains and are considered for release in the workshop meetings.

F: ONE WORD ANSWER

1. A strain which is released for commercial cultivation by a variety release committee.
2. A group of genotypes that are similar in phenotype. After release a strain is called variety.
3. A variety which is developed by the technique of genetic engineering. It differs in one character from the parent variety.
4. A variety which is developed by pure line selection.

5. A variety which is developed by mutation breeding.
6. A variety which is developed by clonal selection, relevant to asexually propagated species.
7. A F_1 population which is used for commercial cultivation.
8. A variety which developed by pedigree breeding method.
9. A variety of self-pollinated crops which is developed by mixing seed of several isogenic lines, related lines or unrelated lines in equal quantity.
10. A variety developed by a farmer.

ANSWERS

MULTIPLE CHOICE QUESTIONS

Sl. No.	*Answer*	*Sl. No.*	*Answer*
1	(b) DDG Crop Science, ICAR	2	Director of Agriculture
3	(d) All of these	4	(d) Any one of these
5	(b) After registration	6	(d) All of these
7	(d) All of these	8	(d) Transgenic variety
9	(d) Transgenic breeding	10	(c) Somatic hybridization

FILL IN THE BLANKS

Sl. No.	*Answer*	*Sl. No.*	*Answer*
1	Cultivar	2	Government of India
3	Registration	4	DDG Crop Science
5	CVRC	6	SVRC
7	Director Agriculture	8	Director Horticulture
9	Variety Release Committee	10	AVT

TRUE OR FALSE

1 to 10 all are true.

APPROPRIATE TERM/ONE WORD ANSWER

Sl. No.	*Answer*	*Sl. No.*	*Answer*
1	Variety	2	Strain
3	Transgenic variety	4	Pure line variety
5	Mutant variety	6	Clonal variety
7	Hybrid variety	8	Pedigree bred variety
9	Multi-line variety	10	Farmer variety

*****

Part II

Seed Production, Certification and Testing

Chapter 6

Principles of Quality Seed Production

A: DEFINITIONS

Breeder's seed: It is a seed or vegetative propagating material which is directly controlled by originating or sponsoring breeder of institution and which provides increases of foundation seeds.

Certified Seed: It is the progeny of foundation, registered or certified seed. This category of seed is distributed to the farmers for commercial crop production.

Foundation seed: It is the progeny of breeder seed which is used for the production of certified seed. Foundation seed is the source of all other certified seed classes, either directly or through registered seed.

Grow out Test: A test which is conducted to assess the genetic purity of a variety during seed production.

Isolation Distance: The separation of the field of a variety from that of another variety of the same crop by a minimum prescribed distance to avoid contamination.

Registered seed: It is the progeny of foundation which is used for the production of certified seed. This category of seed is omitted in India and directly certified seed is produced from the foundation seed.

Rogueing: The process of removal of the off types plants from the field of an improved variety to avoid contamination.

Seed certification: A legal system which ensures availability of high quality seed of improved variety to the farmers.

Volunteer plants: The plants which grow in the field from previous crops.

B: BACKGROUND INFORMATION

1. Seed quality is determined by a number of genetic and physiological characteristics.

2. The genetic component involves differences between two or more genetic lines, while differences between seed lots of a single genetic line comprise the physiological component.

3. The genetic factors that can influence quality include genetic make-up, seed size and bulk density.

4. Deterioration in seed quality may begin at any point in the plant's development stage from fertilization onward.

5. Seed quality depends upon the physical conditions that the mother plant is exposed during its growth stages, as well as harvesting, processing, storage and planting. Temperature, nutrients and other environmental factors also affect seed development and influence seed quality.

6. The maintenance breeding or maintenance of varietal purity is based on certain genetic principles, which are referred to as genetic bases of maintenance breeding.

7. Genetic bases of maintenance breeding include (1) control of seed source, (2) isolation distance, (3) rouging of off types, (4) seed certification and (5) grow out test.

8. For quality seed production, the seed of improved category should be used and obtained from an approved source. Such seed helps in maintaining seed quality in the seed production chain.

9. Four classes of seeds are generally recognized in seed certification chain namely breeder seed, foundation registered and certified. These classes are recognized by Association of official seed certifying agencies [AOSCA].

10. Isolation is essential to prevent pollination from unwanted pollen in the case of cross pollinated and often cross pollinated species to avoid mechanical mixture and chance cross pollination in self-pollinated crops.

11. The isolation distance varies from 3 m in self-pollinated crops like rice, wheat to 200 m in maize, *etc.* It also varies with the type of seeds. In some cases like hybrid maize, the minimum isolation distance may be considerably reduced by planting border rows of pollinator parent and by choosing a large field for seed production.

12. Isolation is also required during harvesting, threshing and handling of seeds to avoid mechanical mixture. The isolation distance is different from crop to crop and from different classes of seeds., *i.e.* certified seeds and foundation seed plots.

13. The off-type plant may arise due to presence of recessive genes in heterozygous condition at the time of release of variety.

14. Off-type plants may also arise from volunteer plants or from seed produced by earlier crop.

15. Mechanical mixtures also lead to presence of off-type plants in a variety.

16. The objective of seed certification is to make available crop seeds, tubers, bulbs, *etc.* of good quality to the farmers. Certification requires two, three or four inspections depending upon the crop species.

17. Varieties being grown for seed production should periodically be tested for genetic purity by grow-out test, to make sure that, seed being maintained in their true form.

18. Grow-out Test is conducted to assess genetic purity during seed production.

C: MULTIPLE CHOICE QUESTIONS

1. Breeder seed is the progeny of:

(a) Nucleus seed (b) Registered seed
(c) Certified seed (d) All of these

2. Foundation seed is the progeny of:

(a) Nucleus seed (b) Breeder seed
(c) Certified seed (d) All of these

3. Certified seed is the progeny of:

(a) Foundation seed (b) Registered seed
(c) Certified seed (d) All of these

4. Which category of seed is omitted in India?

(a) Nucleus seed (b) Registered seed
(c) Certified seed (d) None of these

5. Genetic principles of quality seed production include:

(a) Isolation distance (b) Rogueing
(c) Grow-out Test (d) All of these

6. Which of the following is homozygous and homogeneous?

(a) Mass selected variety (b) Clonal variety
(c) Pure line variety (d) All of these

7. Which of the following is a mixture of several isogenic lines?

(a) Synthetic variety (b) Composite variety
(c) Multiline variety (d) None of these

8. Which of the following is called evolutionary method of breeding?

(a) Pure line selection
(b) Mass selection
(c) Bulk breeding
(d) All of these

9. Which of the following is initial seed of a new variety?

(a) Nucleus seed
(b) Registered seed
(c) Certified seed
(d) All of these

10. Which of the following is used for transfer of male sterility?

(a) Pedigree selection
(b) Bulk breeding
(c) Back cross breeding
(d) All of these

D: FILL IN THE BLANKS

1. Breeder seed is the progeny of ________.
2. Foundation seed is the progeny of ________.
3. Certified seed is the progeny of ________.
4. ________ category of seed is omitted in India.
5. Grow-out test is conducted to assess ________ of a variety.
6. Isolation helps in preventing ________.
7. Rogueing helps in maintaining ________.
8. Rogueing is used to remove ________ plants.
9. Isolation distance is more in ________ species than in self-pollinated ones.
10. Mass selection is based on ________.
11. There is no genetic variation in a ________ variety.
12. A pure line variety is homozygous and ________.
13. In self-pollinated species, a mass selected variety is mixture of several ________.
14. A synthetic variety consists of several ________.
15. A composite variety is a mixture of several ________.

E: MARK TRUE OR FALSE

1. Breeder seed is the progeny of nucleus seed.
2. Foundation seed is the progeny of breeder seed.

3. Certified seed is the progeny of foundation seed.
4. Registered category of seed is omitted in India.
5. Grow-out test is conducted to assess genetic purity of a variety.
6. Isolation helps in preventing outbreeding.
7. Rogueing helps in maintaining varietal purity.
8. Rogueing is used to remove off type plants.
9. Isolation distance is more in cross-pollinated species than in self-pollinated ones.
10. Mass selection is based on phenotype.
11. There is no genetic variation in a pure line variety.
12. A pure line variety is homozygous and homogeneous.
13. In self-pollinated species, a mass selected variety is mixture of several homozygotes.
14. A synthetic variety consists of several homo and heterozygotes.
15. A composite variety is a mixture of several homo and heterozygotes.

F: ONE WORD ANSWER

1. The separation of the field of a variety from that of another variety of the same crop by a minimum prescribed distance to avoid contamination.
2. The process of removal of the off types plants from the field of an improved variety to avoid contamination.
3. A legal system which ensures availability of high quality seed of improved variety to the farmers.
4. A test which is conducted to assess the genetic purity of a variety during seed production.
5. The plants which grow in the field from previous crops.
6. It is a seed or vegetative propagating material which is directly controlled by originating or sponsoring breeder of institution and which provides increases of foundation seeds.
7. It is the progeny of breeder seed which is used for the production of certified seed. Foundation seed is the source of all other certified seed classes, either directly or through registered seed.

8. It is the progeny of foundation which is used for the production of certified seed. This category of seed is omitted in India and directly certified seed is produced from the foundation seed.

9. It is the progeny of foundation, registered or certified seed. This category of seed is distributed to the farmers for commercial crop production.

10. Off type plants in a population.

ANSWERS

MULTIPLE CHOICE QUESTIONS

Sl. No.	*Answer*	*Sl. No.*	*Answer*
1	(a) Nucleus seed	2	(b) Breeder seed
3	(a) Foundation seed	4	(b) Registered seed
5	(d) All of these	6	(c) Pure line variety
7	(c) Multiline variety	8	(c) Bulk breeding
9	(a) Nucleus seed	10	(c) Back cross breeding

FILL IN THE BLANKS

Sl. No.	*Answer*	*Sl. No.*	*Answer*
1	Nucleus seed	2	Breeder seed
3	Foundation seed	4	Registered seed
5	Genetic purity	6	Outcrossing
7	Genetic purity	8	Off type
9	Cross-pollinated	10	Phenotype
11	Pure line	12	Homogeneous
13	Homozygotes	14	Homo and heterozygotes
15	Homo and heterozygotes		

TRUE OR FALSE

1 to 15 all are true.

APPROPRIATE TERM/ONE WORD ANSWER

Sl. No.	*Answer*	*Sl. No.*	*Answer*
1	Isolation Distance	2	Rogueing
3	Seed certification	4	Grow out Test
5	Volunteer plants	6	Breeder's seed
7	Foundation seed	8	Registered seed
9	Certified Seed	10	Off type plants

*****

Chapter 7

Hybrid Seed Production: Field Crops

A: DEFINITIONS

Double Cross Hybrid: The hybrid progeny from a cross between two single crosses, *i.e.* (A x B) x (C x D).

Hybrid: The progeny of a cross between genetically dissimilar parents.

Hybrid Variety: The F_1 population that is used for commercial cultivation.

Isolation Distance: The separation of the field of a variety from that of another variety of the same crop by a minimum prescribed distance to avoid contamination.

Rogueing: The process of removal of the off types plants from the field of an improved variety to avoid contamination.

Single Cross Hybrid: The hybrid progeny from a cross between two inbreds or varieties, *viz.* (A x B).

Three way Cross Hybrid: The hybrid progeny between a single cross and an inbred *viz.* (A x B) x C.

B: BACKGROUND INFORMATION

1. The progeny of a cross between genetically dissimilar parents is called hybrid. Hybrid variety refers to those F_1 populations which are used for commercial cultivation.

2. Hybrid seed production is carried out only for those hybrids which are released either by Central Variety Release Committee or State Variety Release Committee and notified by the Government of India.

3. Hybrid seed is produced by the National Seeds Corporation and State Seed Corporations. These organizations produce hybrid seed of only those hybrids which have been developed by Government Research Institutes

or Agricultural Universities. Private seed companies also produce hybrid seeds of their own hybrids and also of public sector hybrids.

4. Hybrid seed belongs to the certified category and, therefore, certification of hybrid seed is essential.
5. Hybrid seed is produced in two ways, *viz.* by (i) hand emasculation and pollination, and (ii) by using male sterility system. The first method is used for those crops where male sterility system is not available.
6. Either breeder seed or foundation seed of parental lines should be used for production of hybrid seed. Impurity of parental lines leads to rapid deterioration of hybrid.
7. Based on cultivation season, hybrids are classified in two groups, *viz.* kharif season hybrids and rabi season hybrids. Kharif season hybrids are developed in maize, *Sorghum*, pearl millet, pigeon pea, rice and castor. Rabi season hybrids are developed in rapeseed and mustard, safflower and sunflower.
8. Hybrid seed production consists of five important steps such as (i) planting ratio, (ii) isolation distance, (iii) rogueing, and (iv) field inspection.
9. The planting ratio of female and male parents differs from crop to crop. Moreover, in some crops female and male parents are planted in adjacent rows, while in other crops they are planted in separate blocks.
10. Prescribed isolation distance is maintained all around the hybrid seed production plot which varies from crop to crop. However, when hybrid seed production is carried out by hand emasculation and pollination, there is no need of isolation distance such as in cotton.
11. Rogueing is essential in female and male parents for pure hybrid seed production. The rogueing should be done at least, one before flowering and second after flowering.
12. The hybrid seed production plots are inspected by the inspectors of State Seed Certification Agency. Number of inspection differs from crop to crop. Field inspections are carried out to examine (a) isolation distance, (b) off types, (c) weeds, (d) insect and disease incidence, and general crop conditions. All these should be as per prescribed standards.
13. Successful production of hybrid seed depend on several factors such as (i) purity of parental lines, (ii) stability of male sterile line, (iii) restoration capacity of R line, (iv) resistance level of parental lines, (v) Yield level of parental lines, (vi) nicking of female and male lines, and (vii) presence of marker in male sterile line.
14. The genetic purity of parental lines is important in hybrid seed production. It leads to stable performance of hybrid over regions and seasons.

15. Male sterile line should be photo and thermo insensitive. Instability of MS line hampers the hybrid seed production program.

16. The restorer line should be able to completely restore fertility in F_1. Poor restoration capacity hampers hybrid seed production program.

17. Parental line should possess high level of resistance to major insects and disease prevailing in the region where hybrid is commercially cultivated. For example in cotton, in north zone leaf curl virus is a major disease. Hence the parents of hybrid should be resistant to leaf curl virus. Several other such examples can be cited.

18. The yield level of hybrid depends on the yield level of its parents. Hence, yield level of at least one parent of the hybrid should be high. Moreover, parental line should have broad genetic base which will provide wider adaptability to the hybrid.

19. The nicking of flowering period of female and male lines is essential for successful hybrid seed production. If both parents do not flower at one time, the sowing of late flowering parent should be advanced in such a way that their flowering synchronies.

20. When hybrid seed production is carried out using genetic male sterile line, 50 per cent of the male fertile plants have to be removed from the female parent. Identification of male fertile plants is possible only after flower initiation. Presence of marker helps in early detection of male fertile plants during seedling stage.

C: MULTIPLE CHOICE QUESTIUONS

1. Natural outcrossing in maize is more than:

(a) 60 per cent (b) 70 per cent
(c) 80 per cent (d) 90 per cent

2. In maize, an isolation distance required for hybrid seed production is:

(a) 100 m (b) 200 m
(c) 400 m (d) 800 m

3. In pearl millet, cross pollination under natural conditions is more than:

(a) 50 per cent (b) 60 per cent
(c) 80 per cent (d) 90 per cent

4. In pearl millet, an isolation distance required for hybrid seed production is:

(a) 100 m (b) 200 m
(c) 300 m (d) 400 m

5. In Sorghum natural outcrossing ranges from:

(a) 5-10 per cent (b) 10-20 per cent
(c) 10-15 per cent (d) 5-30 per cent

6. In Sorghum, an isolation distance required for hybrid seed production is:

(a) 50 m (b) 100 m
(c) 200 m (d) 300 m

7. In Pigeon pea, an isolation distance required for hybrid seed production is:

(a) 75 m (b) 100 m
(c) 150 m (d) 200 m

8. What is the planting ratio of female and male parents in Maize?

(a) 4 : 1 (b) 4 : 2
(c) 6 : 2 (d) 3 : 1

9. What is the planting ratio of female and male parents in pearl millet?

(a) 3 : 1 (b) 4 : 2
(c) 5 : 2 (d) 3 : 1

10. What is the planting ratio of female and male parents in Sorghum?

(a) 4 : 2 (b) 4 : 1
(c) 6 : 2 (d) 3 : 1

11. What type of hybrids are common in *Sorghum*?

(a) Single cross (b) Three-way cross
(c) Double cross (d) All of these

D: FILL IN THE BLANKS

1. In maize, cross pollination under natural conditions is more than ________.
2. In maize, cross pollination occurs by ________.
3. In maize, an isolation distance of ________meters is required for hybrid seed production.
4. In maize, now ________ cross hybrids are more common.
5. In pearl millet, cross pollination under natural conditions is more than ________.
6. In pearl millet, cross pollination occurs by ________.
7. In pearl millet, an isolation distance of ________ meters is required for hybrid seed production.

8. In pearl millet, ________ cross hybrids are common.
9. In *Sorghum* outcrossing varies from ________.
10. In *Sorghum,* an isolation distance of ________ is required for hybrid seed production.
11. In *Sorghum,* ________ cross hybrids are made.
12. In pigeon pea, cross pollination occurs by honey bees and ________.
13. In pigeon pea, the pollen is heavy and stick hence ________ is not possible.
14. In pigeon pea, an isolation distance of ________ is required for hybrid seed production.
15. Castor is a cross pollinated crop and cross-pollination occurs mainly by ________.
16. In maize, planting ratio of female and male parents is ________.
17. In pearl millet, planting ratio of female and male parents is ________.
18. In *Sorghum,* planting ratio of female and male parents is ________.
19. In sunflower cross pollination occurs by ________.
20. Safflower is a cross pollinated crop and cross pollination occurs mainly by ________.

E: MARK TRUE OR FALSE

1. In maize, cross pollination under natural conditions is more than 90 per cent.
2. Maize pollen is light, hence cross pollination occurs by wind.
3. In maize, an isolation distance of 200 meters is required for hybrid seed production.
4. In maize, both single and double cross hybrids are common.
5. In pearl millet, cross pollination under natural conditions is more than 80 per cent.
6. In pearl millet, cross pollination occurs by wind.
7. In pearl millet, an isolation distance of 200 meters is required for hybrid seed production.
8. In pearl millet, single cross hybrids are made.

9. *Sorghum* is basically self-pollinated crop in which outcrossing varies from 5-30 per cent.

10. In *Sorghum*, an isolation distance of 100 meters is required for hybrid seed production.

11. In *Sorghum*, single cross hybrids are made.

12. In pigeon pea, cross pollination occurs by insects mainly by honey bees and bumble bees.

13. In pigeon pea, the pollen is heavy and stick hence wind pollination is not possible.

14. In pigeon pea, an isolation distance of 100 meters is required for hybrid seed production.

15. Castor is a cross pollinated crop and cross-pollination occurs mainly by wind.

16. In maize, planting ratio of female and male parents is 6 : 2.

17. In pearl millet, planting ratio of female and male parents is 4 : 2.

18. In *Sorghum*, planting ratio of female and male parents is 4 : 2.

19. Sunflower is a highly cross pollinated crop in which cross pollination occurs by wind and insects.

20. Safflower is a cross pollinated crop and cross pollination occurs mainly by wind.

F: ONE WORD ANSWER

1. The progeny of a cross between genetically dissimilar parents.

2. The F_1 population that is used for commercial cultivation.

3. The hybrid progeny from a cross between two inbreds or varieties, *viz.* (A x B).

4. The hybrid progeny between a single cross and an inbred, *viz.* (A x B) x C.

5. The hybrid progeny from a cross between two single crosses., *i.e.* (A x B) x (C x D).

6. The separation of the field of a variety from that of another variety of the same crop by a minimum prescribed distance to avoid contamination.

7. The process of removal of the off types plants from the field of an improved variety to avoid contamination.

ANSWERS

MULTIPLE CHOICE QUESTIONS

Sl. No.	*Answer*	*Sl. No.*	*Answer*
1	(d) 90 per cent	2	(b) 200 m
3	(c) 80 per cent	4	(b) 200 m
5	(d) 5-30 per cent	6	(b) 100 m
7	(b) 100 m	8	(c) 6 : 2
9	(b) 4 : 2	10	(a) 4 : 2
11	(a) Single cross		

FILL IN THE BLANKS

Sl. No.	*Answer*	*Sl. No.*	*Answer*
1	90 per cent	2	Wind
3	200 m	4	Single
5	80 per cent	6	Wind
7	200 m	8	Single
9	5-30 per cent	10	100 m
11	Single	12	Bumble bees
13	Wind	14	100 m
15	Wind	16	6 : 2
17	4 : 2	18	4 : 2
19	Wind	20	Wind

TRUE OR FALSE

1 to 20 all are true.

APPROPRIATE TERM/ONE WORD ANSWER

Sl. No.	*Answer*	*Sl. No.*	*Answer*
1	Hybrid	2	Hybrid Variety
3	Single Cross Hybrid	4	Three way Cross Hybrid
5	Double Cross Hybrid	6	Isolation Distance
7	Rogueing		

*****

Chapter 8

Hybrid Seed Production: Vegetable Crops

A: DEFINITIONS

Double Cross Hybrid: The hybrid progeny from a cross between two single crosses, *i.e.* (A x B) x (C x D).

Hybrid Variety: The F_1 population that is used for commercial cultivation.

Hybrid: The progeny of a cross between genetically dissimilar parents.

Isolation Distance: The separation of the field of a variety from that of another variety of the same crop by a minimum prescribed distance to avoid contamination.

Rogueing: The process of removal of the off types plants from the field of an improved variety to avoid contamination.

Single Cross Hybrid: The hybrid progeny from a cross between two inbreds or varieties, *viz.* (A x B).

Three way Cross Hybrid: The hybrid progeny between a single cross and an inbred, *viz.* (A x B) x C.

B: BACKGROUND INFORMATION

1. Vegetable crops constitute important part of human diet and are important sources of mineral, vitamins and fibers. These constituents are essential in human diet.

2. There are several vegetable crops which are grown in India. However, hybrids are not developed in all the vegetable crops.

3. Important vegetable crops in which hybrids have been developed for commercial cultivation include okra, brinjal, tomato, onion, cabbage,

cauliflower, chillies, carrot, cucumber, bottle-gourd, watermelon and musk melon.

4. Hybrid seed production is carried out only for those hybrids which are released either by Central Variety Release Committee or State Variety Release Committee and notified by the Government of India.

5. Hybrid seed is produced by the National Seeds Corporation and State Seed Corporations. These organizations produce hybrid seed of only those hybrids which have been developed by Government Research Institutes or Agricultural Universities. Private seed companies also produce hybrid seeds of their own hybrids and also of public sector hybrids.

6. Hybrid seed belongs to the certified category and, therefore, certification of hybrid seed is essential.

7. Hybrid seed is produced in two ways, *viz.* by (i) hand emasculation and pollination, and (ii) by using male sterility system. The first method is used for those crops where male sterility system is not available.

8. Either breeder seed or foundation seed of parental lines should be used for production of hybrid seed. Impurity of parental lines leads to rapid deterioration of hybrid.

9. Based on cultivation season, hybrids are classified in two groups, *viz.* kharif season hybrids and rabi season hybrids. Kharif season hybrids are developed in maize, Sorghum, pearl millet, pigeon pea, rice and castor. Rabi season hybrids are developed in rapeseed and mustard, safflower and sunflower.

10. Hybrid seed production consists of five important steps such as (i) planting ratio, (ii) isolation distance, (iii) rogueing, and (iv) field inspection.

11. The planting ratio of female and male parents differs from crop to crop. Moreover, in some crops female and male parents are planted in adjacent rows, while in other crops they are planted in separate blocks.

12. Prescribed isolation distance is maintained all around the hybrid seed production plot which varies from crop to crop. However, when hybrid seed production is carried out by hand emasculation and pollination, there is no need of isolation distance such as in cotton.

13. Rogueing is essential in female and male parents for pure hybrid seed production. The rogueing should be done at least, one before flowering and second after flowering.

14. The hybrid seed production plots are inspected by the inspectors of State Seed Certification Agency. Number of inspection differs from crop to crop. Field inspections are carried out to examine (a) isolation distance,

(b) off types, (c) weeds, (d) insect and disease incidence, and general crop conditions. All these should be as per prescribed standards.

15. Successful production of hybrid seed depend on several factors such as (i) purity of parental lines, (ii) stability of male sterile line, (iii) restoration capacity of R line, (iv) resistance level of parental lines, (v) yield level of parental lines, (vi) nicking of female and male lines, and (vii) presence of marker in male sterile line.

16. The genetic purity of parental lines is important in hybrid seed production. It leads to stable performance of hybrid over regions and seasons.

17. Male sterile line should be photo and thermo insensitive. Instability of MS line hampers the hybrid seed production program.

18. The restorer line should be able to completely restore fertility in F_1. Poor restoration capacity hampers hybrid seed production program.

19. Parental line should possess high level of resistance to major insects and disease prevailing in the region where hybrid is commercially cultivated. For example in cotton, in north zone leaf curl virus is a major disease. Hence the parents of hybrid should be resistant to leaf curl virus. Several other such examples can be cited.

20. The yield level of hybrid depends on the yield level of its parents. Hence, yield level of at least one parent of the hybrid should be high. Moreover, parental line should have broad genetic base which will provide wider adaptability to the hybrid.

21. The nicking of flowering period of female and male lines is essential for successful hybrid seed production. If both parents do not flower at one time, the sowing of late flowering parent should be advanced in such a way that their flowering synchronies.

22. When hybrid seed production is carried out using genetic male sterile line, 50 per cent of the male fertile plants have to be removed from the female parent. Identification of male fertile plants is possible only after flower initiation. Presence of marker helps in early detection of male fertile plants during seedling stage.

C: MULTIPLE CHOICE QUESTIONS

1. In okra, outcrossing is up to:

(a) 10 per cent (b) 15 per cent

(c) 19 per cent (d) 30 per cent

2. In okra, an isolation distance required for hybrid seed production is:

(a) 100 m (b) 200 m

(c) 250 m (d) 300 m

3. What types of hybrids are common in okra?

(a) Single cross (b) Three-way cross
(c) Double cross (d) All of these

4. In brinjal, outcrossing is reported up to:

(a) 15 per cent (b) 20 per cent
(c) 29 per cent (d) 40 per cent

5. In brinjal, an isolation distance required for hybrid seed production is:

(a) 50 m (b) 100 m
(c) 150 m (d) 200 m

6. In pepper, outcrossing is reported up to:

(a) 10 per cent (b) 16.5 per cent
(c) 20 per cent (d) 25 per cent

7. In pepper, an isolation distance required for hybrid seed production is:

(a) 50 m (b) 100 m
(c) 150 m (d) 200 m

8. In onion, an isolation distance required for hybrid seed production is:

(a) 100 m (b) 200 m
(c) 400 m (d) 600 m

9. In cabbage, an isolation distance required for hybrid seed production is:

(a) 200 m (b) 400 m
(c) 800 m (d) 1000 m

10. In bottle gourd, an isolation distance required for hybrid seed production is:

(a) 100 m (b) 200 m
(c) 300 m (d) 400 m

D: FILL IN THE BLANKS

1. Okra is an often cross-pollinated species with outcrossing up to ________.
2. In okra, an isolation distance of ________ is required for hybrid seed production.
3. In okra, ________ cross hybrids are developed.
4. In brinjal outcrossing up to ________ has been reported.
5. In brinjal, Cross-pollination occurs mainly by ________.
6. In brinjal ________ hybrid are developed.

7. In brinjal, an isolation distance of ________ is required for hybrid seed production.
8. In tomato, an isolation distance of ________ is adequate for hybrid seed production.
9. In pepper outcrossing has been reported up to ________.
10. In pepper, cross-pollination occurs mainly by ________.
11. In pepper, an isolation distance of ________ required for hybrid seed production.
12. In onion, an isolation distance of ________ is required for hybrid seed production.
13. In cabbage, an isolation distance of ________ is required for hybrid seed production.
14. In cauliflower, an isolation distance of ________ is required for hybrid seed production.
15. In carrot, an isolation distance of ________ is required for hybrid seed production.
16. In cucumber cross-pollination occurs mainly by ________ and flies.
17. In cucumber, an isolation distance of ________ is required for hybrid seed production.
18. In bottle gourd, an isolation distance of ________ is required for hybrid seed production.
19. In water melon, an isolation distance of ________ is required for hybrid seed production.
20. In musk melon, an isolation distance of ________ is required for hybrid seed production.

E: MARK TRUE OR FALSE

1. Okra is an often cross-pollinated species with outcrossing up to 19 per cent.
2. In okra, the pollen is heavy and sticky and hence wind pollination is not possible.
3. In okra, an isolation distance of 200 meters is required for hybrid seed production.
4. In okra, single cross hybrids are developed.

5. Brinjal is basically a self-pollinated crop in which outcrossing up to 29 per cent has been reported.

6. In brinjal, cross-pollination occurs mainly by honeybees.

7. In brinjal single cross hybrid are developed.

8. In brinjal, an isolation distance of 100 meters is required for hybrid seed production.

9. Tomato is a self-pollinated crop.

10. In tomato, an isolation distance of 25 meters is adequate for hybrid seed production.

11. Pepper is an often cross-pollinated species with outcrossing up to 16.5 per cent. Cross-pollination occurs mainly by honeybees.

12. In pepper, an isolation distance of 200 meters is required for hybrid seed production.

13. Onion is often cross-pollinated crop in which flowers contain nectar glands to attract insects for cross-pollination.

14. In onion, an isolation distance of 400 meters is required for hybrid seed production.

15. Cabbage is a naturally cross-pollinated crop due to presence of sporophytic self-incompatibility. Cross-pollination occurs mainly by honeybees and flies.

16. In cabbage, an isolation distance of 1000 meters is required for hybrid seed production.

17. Cauliflower is a cross-pollinated crop. Cross-pollination occurs mainly by honeybees and flies.

18. In cauliflower, an isolation distance of 1000 meters is required for hybrid seed production.

19. Carrot is a cross-pollinated crop. Cross-pollination occurs mainly by honeybees and flies.

20. In carrot, an isolation distance of 800 meters is required for hybrid seed production.

21. Cucumber is a cross-pollinated crop. Due to presence of sporophytic self-incompatibility. Cross-pollination occurs mainly by honeybees and flies.

22. In cucumber, an isolation distance of 400 meters is required for hybrid seed production.

23. Bottle gourd, water melon and musk melon are cross-pollinated crops, due to their monoecious nature. Cross-pollination occurs mainly by honeybees and flies.

24. In bottle gourd, water melon and musk melon an isolation distance of 400 meters is required for hybrid seed production.

F: ONE WORD ANSWER

1. The progeny of a cross between genetically dissimilar parents.

2. The F_1 population that is used for commercial cultivation.

3. The hybrid progeny from a cross between two inbreds or varieties, *viz.* (A x B).

4. The hybrid progeny between a single cross and an inbred, *viz.* (A x B) x C.

5. The hybrid progeny from a cross between two single crosses., *i.e.* (A x B) x (C x D).

6. The separation of the field of a variety from that of another variety of the same crop by a minimum prescribed distance to avoid contamination.

7. The process of removal of the off types plants from the field of an improved variety to avoid contamination.

ANSWERS

MULTIPLE CHOICE QUESTIONS

Sl. No.	*Answer*	*Sl. No.*	*Answer*
1	(c) 19 per cent	2	(b) 200 m
3	(a) Single cross	4	(c) 29 per cent
5	(b) 100 m	6	(b) 16.5 per cent
7	(d) 200 m	8	(c) 400 m
9	(d) 1000 m	10	(c) 400 m

FILL IN THE BLANKS

Sl. No.	*Answer*	*Sl. No.*	*Answer*
1	19 per cent	2	200 m
3	Single cross	4	29 per cent
5	Honey bees	6	Single cross
7	100m	8	25 m
9	16.5 per cent	10	Honey bees
11	200 m	12	400 m
13	1000 m	14	1000 m
15	800 m	16	Honey bees
17	400 m	18	400 m
19	400 m	20	400 m

TRUE OR FALSE

1 to 24 all are true.

APPROPRIATE TERM/ONE WORD ANSWER

Sl. No.	*Answer*	*Sl. No.*	*Answer*
1	Hybrid	2	Hybrid Variety
3	Single Cross Hybrid	4	Three way Cross Hybrid
5	Double Cross Hybrid	6	Isolation Distance
7	Rogueing		

*****

Chapter 9

Seed Certification

A. DEFINITIONS

Certified seed: The seed produced from foundation, registered or certified seeds.

Grow-Out Test: The test which is carried out to evaluate the seeds for their genuineness to species or varieties or seed-borne infection.

Seed Certification: A legal system which ensures availability of high quality seed of improved variety to the farmers.

Seed Certification Agency: A government organization which is authorized for seed certification. It may have its own seed testing laboratory or it may get its seed samples tested through seed testing laboratories.

Seed certification measures: Various measures which are adopted to ensure quality seed production such as verification of seed source, field inspections, sample inspection, bulk inspection, control plot test and grow-out test.

Seed certification standards: The minimum standards of isolation distance, genetic purity and germination required for the certification of seeds by the certification agencies.

B. BACKGROUND INFORMATION

1. Seed Certification is a legal system which has been designed to maintain and make available to the farmers continuous supply of high quality seeds and propagating materials of released and notified varieties of crop plants.
2. The main objective of the Seed Certification is to ensure the acceptable standards of seed viability, vigour, purity and seed health.
3. The certified seed is annually produced by progressive farmers according to standard seed production practices. To be certified, the seed must meet the prescribed requirements purity, germination and quality.

4. In India, seed certification started in 1963 with the establishment of National Seeds Corporation. Certified seed is produced by registered seed growers.

5. A producer means a person/organisation who grows or distributes certified seed as per procedures and standards of the certification.

6. A legal status to seed certification was given in 1966 with the approval of first Indian Seed Act. In 1968, Seed Rules were formulated.

7. In 1970, the first official State Seed Certifications Agency was established in Maharashtra. In 1974, Karnataka was the first State to establish the Seed Certification Agency as an autonomous body.

8. In 1976, National Seed Project was launched to establish State Seeds Corporations. Now, there are 21 States Seed Certification Agencies in the country which have been established under the Seed Act, 1966.

9. In 2004, the new Seed Act was introduced to replace Seed Act of 1966. In 2005, the Seed Act 2004 came in force in January, 2005.

10. In great majority of the countries in the World, including India, seed certification is voluntary and labelling is compulsory.

11. Seed certification measures include, verification of seed source, field inspections, sample inspection, bulk inspection, control plot test and grow out test.

12. The main purpose of field inspections is to examine five things, *viz.* (i) isolation distance, (ii) off types, (iii) objectionable weeds, (iv) disease and insect incidence, and (v) general crop condition.

13. Field inspections are made by the inspectors of State Seed Certification Agency to examine the suitability of crop for certification.

14. The number of field inspection varies from 2 to 4 depending upon the crop species.

15. When only two inspections are made, one is made during flowering and another before harvesting.

16. In case of three inspections, one is done before flowering, second at the time of flowering and third before harvesting.

17. In case of four inspections, one is done before flowering, second at the time of flowering and third and fourth between flowering and harvesting.

18. Grow-out test is carried out to evaluate the seeds for their genuineness to species or varieties or seed borne infection.

C. MULTIPLE CHOICE QUESTION

1. Requirements of seed certification include:

(a) Improved variety (b) Physical purity
(c) Genetic purity (d) All of these

2. Requirements of seed certification include:

(a) Good germination (b) Good seed health
(c) Proper seed moisture (d) All of these

3. Certified seed is the progeny of:

(a) Nucleus seed (b) Breeder seed
(c) Certified seed (d) All of these

4. Requirements of seed certification include:

(a) Isolation distance (b) Rogueing
(c) Field inspections (d) All of these

5. Seed certification measures include:

(a) Verification of seed source (b) Field inspections
(c) Grow-out test (d) All of these

6. National Seeds Corporation was established in:

(a) 1961 (b) 1963
(c) 1965 (d) 1967

7. The first Indian Seed Act was approved in:

(a) 1963 (b) 1966
(c) 1970 (d) 1972

8. National Seeds Project was launched in:

(a) 1966 (b) 1968
(c) 1970 (d) 1976

9. PPV and FR Act was approved in:

(a) 1961 (b) 1963
(c) 2001 (d) 2005

10. Second Seed Act was approved in:

(a) 1990 (b) 2001
(c) 2004 (d) 2007

D. FILL IN THE BLANKS

1. Seed Certification Agency is a ________ system of seed quality control.
2. The International Crop Improvement Association (ICIA) was formed in ________.
3. ICIA was formed by Scientists from ________.
4. The name of ICIA was changed to Association of Official Seed Certifying Agencies (AOSCA) in ________.
5. National Seeds Corporation was established in ________.
6. The first Seed act in India was approved in ________.
7. The first official State Seed Certifications Agency was established in Maharashtra in ________.
8. Karnataka was the first State to establish the Seed Certification Agency as an autonomous body in ________.
9. National Seed Project was launched in ________.
10. The new Seed Act was introduced in________to replace Seed Act of 1966.
11. The Seed Act 2004 came in force in January________.
12. Now, there are ________ States Seed Certification Agencies in the country.
13. Certified seed is produced by ________ seed growers.
14. Grow-out test is carried out to determine ________ purity of a variety.

E. MARK TRUE OR FALSE

1. Seed Certification Agency is a legal system of seed quality control.
2. International Crop Improvement Association (ICIA) was formed in 1919.
3. ICIA was formed by Scientists from USA and Canada.
4. The name of ICIA was changed to Association of Official Seed Certifying Agencies (AOSCA) in 1969.
5. National Seeds Corporation was established in 1963.
6. The first Indian Seed act was approved in 1966.
7. The first official State Seed Certifications Agency was established in Maharashtra in 1970.

8. Karnataka was the first State to establish the Seed Certification Agency as an autonomous body in 1974.
9. National Seed Project was launched in 1976.
10. The new Seed Act was introduced in 2004 to replace Seed Act of 1966.
11. The Seed Act 2004 came in force in January, 2005.
12. Now, there are 21 States Seed Certification Agencies in the country.
13. Certified seed is produced by registered seed growers.
14. Grow-out test is carried out to determine genetic purity of a variety.

F. ONE WORD ANSWER

1. A legal system which ensures availability of high quality seed of improved variety to the farmers.
2. The seed produced from foundation, registered or certified seeds.
3. The test which is carried out to evaluate the seeds for their genuineness to species or varieties or seed borne infection.
4. A government organization which is authorized for seed certification. It may have its own seed testing laboratory or it may get its seed samples tested through seed testing laboratories.
5. Various measures which are adopted to ensure quality seed production such as verification of seed source, field inspections, sample inspection, bulk inspection, control plot test and grow-out test.
6. The minimum standards of isolation distance, genetic purity and germination required for the certification of seeds by the certification agencies.

ANSWERS

MULTIPLE CHOICE QUESTIONS

Sl. No.	*Answer*	*Sl. No.*	*Answer*
1	(d) All of these	2	(d) All of these
3	(d) All of these	4	(d) All of these
5	(d) All of these	6	(b) 1963
7	(b) 1966	8	(d) 1976
9	(c) 2001	10	(c) 2004

FILL IN THE BLANKS

Sl. No.	*Answer*	*Sl. No.*	*Answer*
1	Legal	2	1919
3	USA and Canada	4	1969
5	1963	6	1966
7	1970	8	1974
9	1976	10	2004
11	January, 2005	12	21
13	Registered	14	Genetic purity

TRUE OR FALSE

1 to 14 all are true.

APPROPRIATE TERM/ONE WORD ANSWER

Sl. No.	*Answer*	*Sl. No.*	*Answer*
1	Seed Certification	2	Certified seed
3	Grow-Out Test	4	Seed Certification Agency
5	Seed certification measures	6	Seed certification standards

*****

Chapter 10

Testing Seed Quality

A. DEFINITIONS

Abnormal Seedlings: Defective seedlings that lack either cotyledons or have stunted root and shoot or their essential structures are so much decayed that they cannot develop into normal plants.

Composite Sample: The seed sample which is formed by mixing all the primary samples of a seed lot.

Dead Seeds: These are nonviable seeds.

Fresh Ungerminated Seeds: These are viable seeds but do not germinate and remain fresh in germination test.

Hard Seeds: The seeds which do not absorb water. Such seed are common in Fabaceae and Malvaceae families. Seed coats of such seeds are impermeable to water.

Inert Matter: Broken or damages seeds, leaf bits, straw, soil particles, stones, *etc.*

Normal Seedlings: Seedlings with normal growth of root and shoot.

Other Seeds: Seeds of other variety of a crop, other crop seeds and weed seeds.

Primary Sample: The small portion of seed drawn from the large seed lot. Several samples are taken from the seed lot to represent the composition of the seed lot.

Pure Seeds: The portion of the seed that belongs to the variety under testing.

Sampling Authority: An officer or inspector of State Seed Certification Agency and authorized to draw seed samples from the seed lots.

Sampling: The process of taking a small portion of seed from the large seed lot.

Seed Viability: The capacity of seeds to germinate under favorable conditions in the absence of dormancy.

Seed Vigour: The general ability of a seed lot to germinate normally over a range of adverse conditions.

Seed-Testing: The laboratory analysis of a seed lot to determine its quality in terms of physical purity, seed health, germination and seed moisture content.

Submitted Sample: The seed sample which is submitted to the seed testing laboratory for testing.

Working Sample: The seed sample which is drawn from the submitted seed sample to carry out the seed testing work.

B. BACKGROUND INFORMATION

1. Seed testing is generally done for improved classes of seed such as breeder seed, foundation seed and certified seed.

2. Seed testing results help in approval or rejection of seed lots for planting purpose. If a seed lot conforms to the prescribed minimum seed standards, it is approved for planting purpose.

3. The seed testing procedure consists of five major steps, *viz.*, sampling, purity analysis, germination test, moisture test and seed health test.

4. Usually seed sampling is done when bagging is completed and the seed is ready for storage or sale. However, sampling can also be done during seed processing. During sampling care should be taken to have true representative of the seed lot.

5. The seed sampling is done by the officer or inspector of State Seed Certification Agency and submitted to the seed testing laboratory. Seed samples submitted by seed inspectors are called official samples.

6. Sometimes, seed samples are also sent by farmers, seed growers and seed farms. Such samples are called service samples.

7. The sample should be true representative of the entire seed lot. Seed sampling includes various aspects such as time of sampling, sampling authority, size of sample, sampling methods, types of sample, *etc.*

8. The sample size varies from crop to crop. The weight of submitted and working samples for purity test of also differs from species to species.

9. There are two types of sampling methods which are commonly used, *viz.*, sampling with hand and sampling with Trier or probe. The hand method is used since long. The Trier method is a scientific method which is widely used for seed sampling. In both the methods efforts should be made to draw true representative samples from the seed lots.

10. The seed samples drawn from the larger seed lots are of four main types, *viz.*, primary sample, composite sample, submitted sample and working sample.

11. Various devices are used to reduce the sample size. The commonly used devices include conical divider [Boerner type], centrifugal divider [Gamet type] and soil divider [Riffle type]. All the three methods are used in seed testing laboratory. The conical divider is widely used. Each divider has some merits and some limitations.

12. Purity analysis is useful in determining the physical purity of the seed. Purity analysis is carried out with working sample.

13. The physical purity of 98 per cent is required in majority of crops for certification purpose. Moreover, the sample should be free from seeds of other variety of the crop and from objectionable weed seeds.

14. The objectionable weed seeds are either wild relatives of a crop or such weed species whose seeds cannot be separated from the seed of main crop

15. Germination test is essential to determine planting value of the seed. Germination test indicates proportion of the seed that will develop into normal plants.

16. Germination also tells a crop grower about a seed lots potential. A seed lot with 80 per cent germination cannot produce more than 80 seedlings/100 seeds. Therefore, if 100 seedlings are needed, a minimum of 125 seeds must be planted (100/.80 = 125).

17. Germination is expressed in percentage. As per ISTA rules, the sample is divided into five components, *viz.* (i) normal seedlings, (ii) abnormal seedlings, (iii) hard seeds, (iv) fresh ungerminated seeds and (v) dead seeds. Generally seed germination is done in the seed germinators. Germination percentage is expressed base on normal seedlings.

18. The speed of germination is an indicator of vigour. The faster germinating seed lot is considered to have more seed vigour. Another characteristic of more vigorous lots is that they store for longer periods of time without loss of germination.

19. Many plant species do well at an alternating 20 and 30°C. For this regime, the chamber is held for 16 hours at 20°C and for the remaining 8 hours of the day at 30°C.

20. The seed viability is tested with a chemical known as tetrazolium chloride. This is a rapid method of testing seed viability.

21. The seed moisture content can be measured in two ways, *viz.*, by oven method and by moisture meter.

22. Moisture meter is widely used and rapid method of measuring seed moisture.

23. Seed health test is conducted to examine presence of pathogens in the seeds as well as insect damage to the seeds. The seed should be free from seed borne diseases and insect damage. This test is conducted by the seed pathologist to ensure good seed health.

C. MULTIPLE CHOICE QUESTION

1. Seed testing is done for:

(a) Breeder seed (b) Foundation seed
(c) Registered seed (d) All of these

2. Seed testing consists of:

(a) Sampling (b) Purity and germination test
(c) Seed moisture and health test (d) All of these

3. Which test is conducted in seed testing?

(a) Physical purity test (b) Germination test
(c) Seed moisture test (d) All of these

4. Seed health test is conducted to examine presence of:

(a) Fungi (b) Bacteria
(c) Insects (d) All of these

5. In a seed sample physical purity test examines proportion of:

(a) Pure seeds (b) Inert matter
(c) Other crop seeds (d) All of these

6. In which crop hirankhuri is an objectionable weed?

(a) Barley (b) Wheat
(c) Maize (d) All of these

7. In which crop wild rice is an objectionable weed?

(a) Barley (b) Wheat
(c) Maize (d) Rice

8. In which crop wild cucurbits are objectionable weed?

(a) Cucurbits (b) Wheat
(c) Maize (d) All of these

9. In which crop sanji is an objectionable weed?

(a) Barley (b) Methi
(c) Maize (d) All of these

10. In which crop wild okra is an objectionable weed?

(a) Cucurbits (b) Wheat

(c) Maize (d) Okra

D. FILL IN THE BLANKS

1. Germination is calculated based on number of ________ seedlings.
2. Tetrazolium is a rapid test of seed ________.
3. Grow-out test is conducted to determine ________ of seed.
4. Generally seed germination is done in the seed ________.
5. Seed sampling is done by seed inspectors of ________.
6. The speed of germination is an indicator of ________.
7. Germination test is essential to determine planting ________ of the seed.
8. Hirankhuri is an objectionable weed in ________.
9. Wild rice is an objectionable weed in ________.
10. Wild cucurbits are objectionable weed in ________.
11. Sanji is an objectionable weed in ________.
12. Wild okra is an objectionable weed in ________.

E. MARK TRUE OR FALSE

1. Germination is calculated based on number of normal seedlings.
2. Tetrazolium is a rapid test of seed viability.
3. Grow-out test is conducted to determine genetic purity of seed.
4. Generally seed germination is done in the seed germinators.
5. Seed sampling is done by seed inspectors of SSCA.
6. The speed of germination is an indicator of vigour.
7. Germination test is essential to determine planting value of the seed.
8. Hirankhuri is an objectionable weed in wheat.
9. Wild rice is an objectionable weed in rice.
10. Wild cucurbits are objectionable weed in cucurbits.

11. Sanji is an objectionable weed in methi.

12. Wild okra is an objectionable weed in okra.

F. ONE WORD ANSWER

1. The laboratory analysis of a seed lot to determine its quality in terms of physical purity, seed health, germination and seed moisture content.

2. The capacity of seeds to germinate under favorable conditions in the absence of dormancy.

3. The process of taking a small portion of seed from the large seed lot.

4. An officer or inspector of State Seed Certification Agency and authorized to draw seed samples from the seed lots.

5. The small portion of seed drawn from the large seed lot. Several samples are taken from the seed lot to represent the composition of the seed lot.

6. The seed sample which is formed by mixing all the primary samples of a seed lot.

7. The seed sample which is submitted to the seed testing laboratory for testing.

8. The seed sample which is drawn from the submitted seed sample to carry out the seed testing work.

9. The portion of the seed that belongs to the variety under testing.

10. Seeds of other variety of a crop, other crop seeds and weed seeds.

11. Broken or damages seeds, leaf bits, straw, soil particles, stones, *etc.*

12. Seedlings with normal growth of root and shoot.

13. Defective seedlings that lack either cotyledons or have stunted root and shoot or their essential structures are so much decayed that they cannot develop into normal plants.

14. The seeds which do not absorb water. Such seed are common in *Fabaceae* and *Malvaceae* family. Seed coats of such seeds are impermeable to water.

15. These are viable seeds but do not germinate and remain fresh in germination test.

16. These are nonviable seeds.

17. The general ability of a seed lot to germinate normally over a range of adverse conditions.

ANSWERS

MULTIPLE CHOICE QUESTIONS

Sl. No.	*Answer*	*Sl. No.*	*Answer*
1	(d) All of these	2	(d) All of these
3	(d) All of these	4	(d) All of these
5	(d) All of these	6	(b) Wheat
7	(d) Rice	8	(a) Cucurbits
9	(b) Methi	10	(d) Okra

FILL IN THE BLANKS

Sl. No.	*Answer*	*Sl. No.*	*Answer*
1	Normal	2	Viability
3	Genetic purity	4	Germinator
5	SSCA	6	Vigor
7	Value	8	Wheat
9	Rice	10	Cucurbits
11	Methi	12	Okra

TRUE OR FALSE

1 to 12 all are true.

APPROPRIATE TERM/ONE WORD ANSWER

Sl. No.	*Answer*	*Sl. No.*	*Answer*
1	Seed Testing	2	Seed Viability
3	Sampling	4	Sampling Authority
5	Primary Sample	6	Composite Sample
7	Submitted Sample	8	Working Sample
9	Pure Seeds	10	Other Seeds
11	Inert Matter	12	Normal Seedlings
13	Abnormal Seedlings	14	Hard Seeds
15	Fresh Ungerminated Seeds	16	Dead Seeds
17	Seed vigour		

*****

Chapter 11

Maintenance Breeding

A. DEFINITIONS

Maintenance breeding: An area of plant breeding which deals with principles and methods of breeder seed production; also known as varietal maintenance technology.

Model Bulk Selection: Selecting large number of single plants [300-500] from the breeder seed plot of a variety, examining these plants under laboratory conditions for the characteristics of the variety recorded at the time of its release and selecting those plants which conform to the characteristics of a variety and their seed is bulked to grow next generation.

Negative Mass Selection: Removal of undesirable plants [off types] from the field of a variety and mixing the seed of remaining plants to grow the next generation.

Plant to Row Method: This method of maintenance breeding is based on progeny test. It is also known as progeny row selection. Progeny of single plant selections are planted, evaluated and progeny with true to the type are selected and their seed is mixed to grow the next generation.

Positive Mass Selection: Selection of desirable plants [true to type] from the field of a variety and mixing the seed of individual selected plants to grow the next generation.

B. BACKGROUND INFORMATION

1. There are three important areas or activities of plant breeding, *viz.* (i) varietal development, (ii) seed multiplication, and (iii) varietal maintenance or maintenance breeding.

2. Maintenance breeding is used to enhance the life span of released and notified varieties and commercial hybrids. Maintenance breeding is also known as varietal maintenance technology.

3. The main purpose of maintenance breeding is to provide fresh breeder seed every year which is used to start another cycle of seed multiplication.

4. Maintenance breeding prevents varieties from genetic deterioration and thus enhances life span of a variety. It provides opportunities to study the efficacy of various maintenance procedures. It helps in purification of improved varieties and parental lines of commercial hybrids. It leads to quality seed production in the seed production chain. Through maintenance breeding the originality of the variety is maintained.

5. It has to maintain the original performance of a variety recorded at the time of its release. Thus, it prevents varietal deterioration. All the characters of a variety recorded at the time of its release are maintained during breeder seed production.

6. Nucleus or breeder seed is used as base material for starting maintenance breeding program. Sometimes, foundation seed may also be used to start maintenance breeding program.

7. The maintenance breeding is restricted to the breeder seed production of released and notified varieties and parents of commercially cultivated hybrids.

8. In the maintenance breeding, genetic purity, physical purity, germination and seed health are the main criteria taken in to account.

9. Various breeding procedures or techniques are used for maintenance breeding. Important breeding procedures used for maintenance breeding include positive mass selection, negative mass selection, single plant progeny selection, *etc.* The selection applied for maintenance breeding purposes does not ensure genetic gain in the genetic value of a variety.

10. Four breeding methods, *viz.*, positive mass selection, negative mass selection, model bulk selection, and plant to row selection are commonly used for maintaining varietal purity.

11. Positive mass selection is widely used for varietal purification in various field crops. In cotton, this method was first used by Menbane in Texas, USA. Now this method is widely used for maintenance of varietal purity in cotton.

12. Model bulk selection was first used by Manning (1955) in cotton. Now this method is widely used for maintenance breeding.

13. Plant to row selection method of maintenance breeding is based on progeny test. It is also known as progeny row selection.

14. For maintenance breeding most of the testing procedures are based on phenotypic performance which is not always an indication of superior genotypes.

C. MULTIPLE CHOICE QUESTIONS

1. In which method off type plants are rejected in varietal maintenance?

(a) Negative mass selection (b) Positive mass selection
(c) Progeny selection (d) Recurrent selection

2. In which method true to type plants are selected?

(a) Negative mass selection (b) Positive mass selection
(c) Progeny selection (d) Recurrent selection

3. What are methods of maintenance breeding?

(a) Mass selection (b) Progeny selection
(c) Model bulk selection (d) All of these

4. Model bulk selection was first used by Manning (1955) in:

(a) Maize (b) Cotton
(c) Soybean (d) Brassica

5. Positive mass selection in cotton was first used by:

(a) Manning (b) Menbane
(c) Allard (d) Briggs

6. In a variety, maintenance breeding does not ensure:

(a) Genetic purity (b) Physical purity
(c) Genetic gain (d) All of these

7. Maintenance breeding leads to regular fresh supply of:

(a) Breeder seed (b) Foundation seed
(c) Certified seed (d) Registered seed

8. In a variety, maintenance breeding enhances:

(a) Yield (b) Quality
(c) Life span (d) All of these

9. In a variety, maintenance breeding retains its:

(a) Originality (b) Genetic purity
(c) Physical purity (d) All of these

10. Which of the following methods is based on progeny test?

(a) Negative mass selection (b) Positive mass selection
(c) Progeny selection (d) Recurrent selection

D. FILL IN THE BLANKS

1. Maintenance breeding is also known as ________ technology.
2. Off type plants are rejected in ________ mass selection.
3. True to type plants are selected in ________mass selection.
4. Plant to row selection is also known as ________ selection.
5. Maintenance breeding ensures continuous supply of ________.
6. Mass selection is based on ________performance.
7. Plant row selection is based on ________.
8. Maintenance breeding retains the ________ of the variety.
9. Maintenance breeding prevents varieties from ________.
10. Maintenance breeding enhances ________ of a variety.
11. Model bulk selection was first used by Manning (1955) in ________.
12. Maintenance breeding does not ensure ________ in the genetic value of a variety.

E. TRUE OR FALSE

1. Maintenance breeding is also known as varietal maintenance technology.
2. Off type plants are rejected in negative mass selection.
3. True to type plants are selected in positive mass selection.
4. Plant to row selection is also known as progeny row selection.
5. Maintenance breeding ensures continuous supply of breeder seed.
6. Mass selection is based on phenotypic performance.
7. Plant row selection is based on progeny test.
8. Maintenance breeding retains the originality of the variety.
9. Maintenance breeding prevents varieties from genetic deterioration.
10. Maintenance breeding enhances life span of a variety.
11. Model bulk selection was first used by Manning (1955) in cotton.
12. Maintenance breeding does not ensure genetic gain in the genetic value of a variety.

F. ONE WORD ANSWER

1. An area of plant breeding which deals with principles and methods of breeder seed production.

2. Selection of desirable plants [true to type] from the field of a variety and mixing the seed of individual selected plants to grow the next generation.

3. Removal of undesirable plants [off types] from the field of a variety and mixing the seed of remaining plants to grow the next generation.

4. Selecting large number of single plants [300-500] from the breeder seed plot of a variety, examining these plants under laboratory conditions for the characteristics of the variety recorded at the time of its release and selecting those plants which conform to the characteristics of a variety and their seed is bulked to grow nest generation.

5. Progeny of single plant selections are planted, evaluated and progeny with true to the type are selected and their seed is mixed to grow the nest generation.

ANSWERS

MULTIPLE CHOICE QUESTIONS

Sl. No.	*Answer*	*Sl. No.*	*Answer*
1	(a) Negative mass selection	2	(b) Positive mass selection
3	(d) All of these	4	(b) Cotton
5	(b) Menbane	6	(c) Genetic gain
7	(a) Breeder seed	8	(c) Life span
9	(d) All of these	10	(c) Progeny selection

FILL IN THE BLANKS

Sl. No.	*Answer*	*Sl. No.*	*Answer*
1	Varietal maintenance	2	Negative
3	Positive	4	Progeny row
5	Breeder	6	Phenotypic
7	Progeny test	8	Originality
9	Genetic deterioration	10	Life span
11	Cotton	12	Genetic gain

TRUE OR FALSE

1 to 12 all are true

APPROPRIATE TERM/ONE WORD ANSWER

Sl. No.	*Answer*	*Sl. No.*	*Answer*
1	Maintenance breeding	2	Positive Mass Selection
3	Negative Mass Selection	4	Model Bulk Selection
5	Plant to Row Method		

*****

Part III

Miscellaneous Topics

Chapter 12

Principles of DUS Testing

A. DEFINITIONS

Candidate variety: A variety to be registered under Plant Variety Protection Act.

DUS Testing: Evaluation of plant varieties in terms of distinctiveness, uniformity and stability.

Distinctiveness: The new variety must be clearly distinguishable by one or more characters from any other variety whose existence is a matter of common knowledge at the time when the protection is applied for.

Extant Variety: All released, notified and unprotected varieties.

Example Variety: A variety that is used for comparison for a particular character.

Farmers' Rights: Legal rights provided to the farmers to save, use, sow, replant, exchange, share or sell his farm produce including seed of a variety protected under Plant Variety Protection Act.

Farmers' Variety: A variety that has been developed by a farmer and used for commercial cultivation for several years.

Novelty: It refers to newness of a variety. The variety should be new one and should not have been commercially cultivated for more than one year before granting protection under PVP Act.

Plant Breeders' Rights: Legal rights granted to the plant breeder for a new plant variety. It provides rights for commercial seed production, marketing, export and import, authorization and to prevent infringement.

Reference Variety: All released and notified extant varieties of common knowledge which are in seed production chain.

Stability: The variety is deemed to be stable if its relevant characteristics remain unchanged after repeated propagation.

Uniformity: The variety should be sufficiently uniform in its relevant characteristics. Relevant characteristics include all those used as a basis for distinctiveness or included in the variety description established at the date of grant of protection of that variety.

B. BACKGROUND INFORMATION

1. The main objectives of the PPV and FR Act are: (i) registration of plant varieties, (ii) characterization, cataloguing and documentation of registered varieties, (iii) ensuring that seeds of all registered varieties are made available to farmers, and (iv) maintenance of National Register of Plant varieties.

2. The testing of DUS characters is useful in for (i) identification of plant varieties, (ii) registration and protection of plant varieties under Protection of Plant Varieties and Farmers Rights Act 2001, (iii) varietal information system and classification of varieties in to different groups, and (iv) creating data base for plant genetic resources.

3. DUS testing is essential for registration of varieties with Protection of Plant Varieties and Farmers Rights Act, 2001. Any protected variety must be given a single and distinct denomination [name] in accordance with the regulations. Registration of any new plant variety under the PPV FR is voluntary.

4. Any person who markets the produce of this protected variety must use the same denomination even if the period of protection is expired which is again a valuable part of Breeders' rights.

5. In connection with plant variety protection (PVP) Act, various terms such as extant variety, candidate variety, reference variety, example variety and farmers' variety are frequently used. Hence, knowledge of these terms is essential.

6. The PPV and FR Act, 2001 provides that protection shall only be granted after examination of the variety for Distinctness, Uniformity and Stability and Novelty.

7. The first three requirements were as per UPOV Act, 1978 and are known as DUS. The fourth was included in UPOV Act, 1991 and known as NDUS.

8. The period of protection varies with plant species. For field crops, the maximum period of protection is 15 year, whereas for forest trees, fruit trees, ornamental trees, shrubs and vines it is 18 year.

9. An application is to be submitted in the prescribed form with varietal description bringing out Distinctness, Uniformity, Stability and Novelty.

10. Complete passport data of parental lines, their geographical origin is to be furnished.

11. An affidavit declaring that the variety does not contain terminator gene sequence and it was lawfully acquired.

12. Prescribed quantity of seed of the variety and parental lines (for hybrids) is to be deposited with PPV and FR Act Authority. Specified fee for registration and conducting DUS testing is to be paid.

13. DUS test is conducted by adopting prescribe statistical design. The minimum duration of tests should normally be two independent similar growing seasons at two locations.

14. DUS testing is carried at the research centers approved for such purpose. In cotton, four DUS testing centres, *viz.* Nagpur, Coimbatore, Dharwad and Hisar have been identified.

15. Observations are recorded on three types of traits, *viz.* (i) grouping characters, and (ii) essential characters, and (iii) Additional characters.

16. Grouping characters are highly heritable and remain unchanged under varying environmental conditions Such characters are suitable for grouping of varieties into different classes. In cotton, grouping characters include leaf shape, petal color, boll shape and fibre length.

C. MULTIPLE CHOICE QUESTIONS

1. As per UPOV Act 1978, requirements for registration are:

(a) Distinctiveness (b) Uniformity

(c) Stability (d) All of these

2. All released, notified and unprotected varieties are called:

(a) Extant varieties (b) Candidate variety

(c) Reference variety (d) Example variety

3. A variety to be registered under Protection of Plant Variety and FR Act is called:

(a) Extant varieties (b) Candidate variety

(c) Reference variety (d) Example variety

4. All released and notified extant varieties of common knowledge which are in seed production chain refer to:

(a) Extant varieties (b) Candidate variety

(c) Reference variety (d) Example variety

5. A variety that is used for comparison for a particular character is called:

(a) Extant varieties (b) Candidate variety

(c) Reference variety (d) Example variety

6. Characterization refers to recording observations on:.

(a) All characters (b) Heritable characters

(c) Non-heritable characters (d) Diseases

7. Evaluation refers to recording observations on:.

(a) All characters (b) Heritable characters

(c) Non-heritable characters (d) Diseases

8. Screening refers to recording observations on:.

(a) All characters (b) Heritable characters

(c) Non-heritable characters (d) Single character

9. The maximum period of protection for field crops is:

(a) 10 years (b) 12 years

(c) 15 years (d) 20 years

10. The maximum period of protection for forest trees, fruit trees, ornamental trees, shrubs and vines is:

(a) 15 years (b) 18 years

(c) 20 years (d) 25 years

D. FILL IN THE BLANKS

1. All released, notified and unprotected varieties are called ________.
2. A variety to be registered under Plant Variety Protection is called ________.
3. All released and notified extant varieties of common knowledge which are in seed production chain refer to ________.
4. A variety that is used for comparison for a particular character ________.
5. Newness of a variety is called ________.
6. Legal rights granted to the plant breeder for a new plant variety are called ________.
7. Recording observations on highly heritable characters is called ________.
8. Recording observations on several characters refers to ________.
9. Recording observation on single character refers to ________.

10. Closeness on an observation to the average value is called ________.

E. MARK TRUE OR FALSE

1. All released, notified and unprotected varieties are called extant varieties.
2. A variety to be registered under Protection of Plant Variety and FR Act is called candidate variety.
3. All released and notified extant varieties of common knowledge which are in seed production chain refer to reference variety.
4. A variety that is used for comparison for a particular character is called example variety.
5. Newness of a variety is called novelty.
6. Legal rights granted to the plant breeder for a new plant variety are called plant breeders' rights.
7. Recording observations on highly heritable characters is called characterization.
8. Recording observations on several characters refers to evaluation.
9. Recording observation on single character refers to screening.
10. Closeness on an observation to the average value is called precision.

F. ONE WORD ANSWER

1. Evaluation of plant varieties in terms of distinctiveness, uniformity and stability.
2. All released, notified and unprotected varieties.
3. A variety to be registered under Plant Variety Protection Act.
4. All released and notified extant varieties of common knowledge which are in seed production chain.
5. A variety that is used for comparison for a particular character.
6. A variety that has been developed by a farmer and used for commercial cultivation for several years.
7. The new variety must be clearly distinguishable by one or more characters from any other variety whose existence is a matter of common knowledge at the time when the protection is applied for.

8. The variety should be sufficiently uniform in its relevant characteristics. Relevant characteristics include all those used as a basis for distinctiveness or included in the variety description established at the date of grant of protection of that variety.

9. The variety is deemed to be stable if its relevant characteristics remain unchanged after repeated propagation.

10. It refers to newness of a variety. The variety should be new one and should not have been commercially cultivated for more than one year before granting protection under PVP Act.

11. Legal rights granted to the plant breeder for a new plant variety. It provides rights for commercial seed production, marketing, export and import, authorization and to prevent infringement.

12. Legal rights provided to the farmers to save, use, sow, replant, exchange, share or sell his farm produce including seed of a variety protected under Plant Variety Protection Act.

ANSWERS

MULTIPLE CHOICE QUESTIONS

Sl. No.	*Answer*	*Sl. No.*	*Answer*
1	(d) All of these	2	(a) Extant varieties
3	(b) Candidate variety	4	(c) Reference variety
5	(d) Example variety	6	(b) Heritable characters
7	(a) All characters	8	(d) Single character
9	(c) 15 years	10	(b) 18 years

FILL IN THE BLANKS

Sl. No.	*Answer*	*Sl. No.*	*Answer*
1	Extant varieties	2	Candidate variety
3	Reference variety	4	Example variety
5	Novelty	6	Plant breeders' rights
7	Characterization	8	Evaluation
9	Screening	10	Precision

TRUE OR FALSE

1 to 10 all are true

APPROPRIATE TERM/ONE WORD ANSWER

Sl. No.	*Answer*	*Sl. No.*	*Answer*
1	DUS Testing	2	Extant Variety
3	Candidate Variety	4	Reference Variety
5	Example Variety	6	Farmers' Variety
7	Distinctiveness	8	Uniformity
9	Stability	10	Novelty
11	Plant Breeders' Rights	12	Farmers' Rights

*****

Chapter 13

Protection of Plant Varieties and Farmers' Rights Act 2001

A. DEFINITIONS

Candidate Variety: A variety that has to be protected.

Distinctiveness:. The variety should be clearly distinguishable from previously available varieties at least for one characteristic.

Example variety: A variety that is used for comparison of a particular character.

Extant variety: All released and notified varieties which have not been protected.

Farmers' Variety: A variety that has been developed by a farmer and used for commercial cultivation for several years.

Novelty: Newness of a variety. A variety which has not been grown for more than one year prior the application for registration.

Reference variety: All released and notified extant varieties that are under seed production chain.

Stability: The same performance of a variety after repeated reproduction or propagation.

Uniformity: The same type of population of a variety.

B. BACKGROUND INFORMATION

1. The Protection of Plant Varieties and Farmers' Rights Act 2001 was approved by the government of India in 2004 but came into force with effect from January 2005.

2. The main objective of this act is to encourage development of new plant varieties and promote development of seed industry in India.

3. The members of the Authority include (i) Agricultural Commissioner-GOI, (ii) DDG Crop Sciences-ICAR, (iii) Joint Secretary Seeds-GOI, (iv) Horticulture Commissioner-GOI, (v) Director, NBPGR; (vi) one member each [not below the rank of Joint Secretary] from DBT, (vii) Ministry of Environment and forests-GOI, (viii) Ministry of Law Justice and Company Affairs- GOI; (ix) one representative each [nominated by central government] from farmers, (x) tribal organization, (xi) seed industry, (xii) an agricultural university, (xiii) and women organization; and (xi**v)** two nominated representative of state government.

4. The headquarters of Protection of Plant Varieties and Farmers' Rights Authority is located in Pusa Campus [National Agricultural Science Complex], New Delhi.

5. Various types of crop cultivars such as pure line, multiline varieties, open pollinated, synthetic and composite, single, three-way and double cross hybrids and parents of hybrids, trees, fruits, ornamental plants and vines, *etc.* can be protected under PPV and FR Act.

6. The chairperson of PPV and FR act is called Authority. The PPV and FR Act Authority establishes a Plant Varieties Registry, which maintains a National Register of Plant Varieties.

7. A new variety will be registered under the act if it conforms to the criteria of novelty, distinctiveness, uniformity and stability.

8. The PPV and FR Authority has accepted UPOV Act 1978. According to this Act, the period of protection is 15 years for annual and biennial crops and 18 years for perennial plants such as trees and vines. The registration is renewable after its expiry for a similar term of duration.

9. A farmer is entitled to save, use, sow, resow, exchange, share or sell his farm produce including seed of a variety protected under the Act in the same manner before this Act came into force. He cannot sell branded seed of a variety protected under the Act.

10. A certificate of registration for a variety issued by this act shall confer an exclusive right on the breeder or his successor, his agent or licensee, to produce, sell, market, distribute, import or export seed of his variety.

11. The breeder has right to authorize any other person for production and marketing of his variety. If a breeder of a propagating material of a variety registered under the Act sells his product to a farmer, he has to disclose the expected performance under given conditions. If the propagating material fails to perform, the farmer can claim compensation in the prescribed manner before the Protection of Plant Varieties and Farmers' Rights Authority.

12. Those techniques which are harmful to animals and the environment such as terminator technology and traitor technology are excluded and not allowed by the Protection of Plant Varieties and farmers' Rights Act 2001.

13. Penalty for applying false denomination to a variety is imprisonment up to two years and/or a fine between Rs 50,000 and Rs five lakh.

14. Penalty for falsely representing a variety as registered is imprisonment up to three years and/or a fine between Rs one lakh and Rs five lakh or both.

15. Penalty for subsequent offence is imprisonment up to three years and/ or a fine between Rs two lakh and Rs 20 lakh.

16. It provides an effective system for protection of plant varieties, farmers' rights and plant breeders' rights.

17. It recognizes the contribution of farmers in conserving, improving and making available plant genetic resources for development of new plant varieties.

18. It provides rights to farmers to save, use, sow, resow, exchange, share or sell his farm produce including seed of a variety protected by PPV and FR Act.

19. It confers an exclusive right on the breeder or his successor, his agent or licensee, to produce, sell, market, distribute, import or export seed of his varietyThus it gives recognition to breeders for their contribution and encourage them to breed still better varieties.

20. It promotes the growth of seed industry in the country to ensure availability of high quality seeds to the farmers.

21. It involves representatives of farmers, women and seed industry in its legislative structure who will look to the interest of these sectors.

22. The Act has provision for compensation. If the seed gives poor performance than the declared by the breeder, farmer can claim compensation in the prescribed manner before the Protection of Plant Varieties and Farmers' Rights Authority.

23. The Act has provision for rewarding farmers for their role in conserving plant genetic resources. Such expenditure can be met out from the gene fund available with the PPV and FR Act Authority.

24. The act has provision of benefit sharing. The farmers can get share from the benefits arising from cultivation of farmers varieties and knowledge.

25. The act has provision to impose penalties to curb unhealthy practices.
26. The PPV and FR Act will promote cultivation of new high yielding varieties which are highly uniform. This will result in reduction in genetic diversity leading to narrow genetic base and danger of uniformity.
27. This will encourage monopoly of some breeders or seed companies for genetic material with special traits

C. MULTIPLE CHOICE QUESTIONS

1. PPV and FR Act was proposed in:

(a) 2001 (b) 2003
(c) 2005 (d) 2006

2. PPV and FR Act was approved in:

(a) 2001 (b) 2004
(c) 2005 (d) 2006

3. PPV and FR Act came in to force in:

(a) 2001 (b) 2003
(c) 2005 (d) 2006

4. The headquarters of PPV and FR authority is located at:

(a) Chennai (b) New Delhi
(c) Mumbai (d) Kolkata

5. The maximum period of protection for field crops is:

(a) 10 years (b) 12 years
(c) 15 years (d) 20 years

6. The maximum period of protection for trees and vines is:

(a) 15 years (b) 18 years
(c) 20 years (d) 25 years

7. Penalty for applying false denomination to a variety is imprisonment up to:

(a) 3 months (b) 6 months
(c) 1 year (d) 2 years

8. Penalty for falsely representing a variety as registered is imprisonment up to:

(a) 6 months (b) 12 months
(c) 3 years (d) 5 years

9. PPV and FR Act, in its central committee, involves representatives of:

(a) Farmers (b) Women

(c) Seed Industry (d) All of these

10. Compensation for poor performance of a variety than declared can be claimed from:

(a) District court (b) Consumer court

(c) High court (d) PPV and FR Act authority

D. FILL IN THE BLANKS

1. PPV and FR Act was proposed in ________.
2. A new variety can be registered with ________ Authority.
3. The headquarters of PPV and FR Act authority is located at ________.
4. PPV and FR Act was approved by the central government in ________.
5. PPV and FR Act came in to force from ________.
6. PPV and FR Act authority has accepted UPOV Act ________.
7. The maximum period of protection for new varieties of field crops is ________ years.
8. The maximum period of protection for new varieties of trees and vines is ________years.
9. No fee is charged if a variety is registered by a ________.
10. Farmers cannot sale seed of protected variety under ________ name.
11. PPV and FR Act does not permit use of ________ technology.
12. PPV and FR Act also does not allow use of ________technology.
13. Primitive cultivars are known as ________.
14. Use of high yielding varieties will lead to reduction in ________.
15. Farmers can be rewarded for their contribution from gene ________.

E. MARK TRUE OR FALSE

1. PPV and FR Act was proposed in 2001.
2. A new variety can be registered with PPV and FR Authority.
3. The headquarters of PPV and FR Act authority is located at, New Delhi.

4. PPV and FR Act was approved by the central government in 2004.
5. PPV and FR Act came in to force from January, 2005.
6. PPV and FR Act authority has accepted UPOV Act, 1978.
7. The maximum period of protection for new varieties of field crops is 15 years.
8. The maximum period of protection for new varieties of trees and vines is 18 years.
9. No fee is charged if a variety is registered by a Farmer.
10. Farmers cannot sale seed of protected variety under brand name.
11. PPV and FR Act does not permit use of terminator technology.
12. PPV and FR Act also does not allow use of traitor technology.
13. Primitive cultivars are known as land races.
14. Use of high yielding varieties will lead to reduction in biodiversity.
15. Farmers can be rewarded for their contribution from gene fund.

F. ONE WORD ANSWER

1. A variety which has not been grown for more than one year prior the application for registration.
2. The variety should be clearly distinguishable from previously available varieties at least for one characteristic.
3. The same type of population of a variety.
4. The same performance of a variety after repeated reproduction or propagation.
5. All released and notified varieties which have not been protected.
6. A variety that has to be protected.
7. All released and notified extant varieties that are under seed production chain.
8. A variety that is used for comparison of a particular character.
9. A variety that has been developed by a farmer and used for commercial cultivation for several years.

ANSWERS

MULTIPLE CHOICE QUESTIONS

Sl. No.	*Answer*	*Sl. No.*	*Answer*
1	(a) 2001	2	(b) 2004
3	(c) 2005	4	(b) New Delhi
5	(c) 15 years	6	(b) 18 years
7	(d) 2 years	8	(c) 3 year
9	(d) All of these	10	(d) PPV and FR Act authority

FILL IN THE BLANKS

Sl. No.	*Answer*	*Sl. No.*	*Answer*
1	2001	2	PPV and FR
3	New Delhi	4	2004
5	January, 2005	6	1978
7	15	8	18
9	Farmer	10	Brand
11	Terminator	12	Traitor
13	Land races	14	Biodiversity
15	Gene fund		

TRUE OR FALSE

1 to 15 all are true

APPROPRIATE TERM/ONE WORD ANSWER

Sl. No.	*Answer*	*Sl. No.*	*Answer*
1	Novelty	2	Distinctiveness
3	Uniformity	4	Stability
5	Extant variety	6	Candidate Variety
7	Reference variety	8	Example variety
9	Farmers' Variety		

****

Chapter 14

Intellectual Property Rights

A. DEFINITIONS

Copyrights: The exclusive rights to reproduce, sell and distribute a work, prepare derivative work and display the work publicly.

Database Rights: A 'database right' is an intellectual property right given to a computer database.

Domain Name: The names and words that companies designate for their registered Internet Web site addresses.

Exclusive Rights: The legal privileges that only a copyright owner has with respect to their copyrighted work.

Farmers' Rights: The legal rights provided to farmers to save, use, sow, replant, exchange, share or sell his farm produce including seed of a variety protected under Plant Variety Protection Act.

Geographical Indication: A geographical indication is a sign used on goods that have a specific geographical origin and often possess qualities or a reputation that are due to that place of origin.

Immovable Property: Fixed types of properties which cannot move from one place to other. Such properties include land, buildings [house, flat, bungalow, farm house] and gardens.

Industrial Design: Any original shape, picture, or some combination applied to a useful article of manufacture.

Intellectual Property Rights: The legal rights provided to an inventor to derive economic benefits from his invention/innovation.

Intellectual Property: The product/process/idea which is outcome of the brain of a person and can be used on commercial scale for benefit of human kind.

Mask Work: A mask work is a two or three-dimensional layout of an integrated circuit.

Moral Rights: Moral rights are a special extension to copyright that gives the originator of a copyrighted work rights over its use.

Movable Property: The property which can be easily shifted from one place to other. Such property includes animals, farm machines, furniture, fixtures, gold, silver, diamond, and money.

Patent: A document granting an inventor sole rights to an invention. It is an official document which grants sole rights to the inventor for manufacturing and marketing his product/process/invention to derive benefits.

Plant Breeders' Rights: Plant breeders' rights, also known as plant variety rights (PVR), are intellectual property rights granted to the breeder of a new variety of plant.

Property: The wealth or valuable things earned by a person.

Sui Generis Rights: Sui-generis refers to things of their own kind or things with unique characteristics.

Supplementary Protection Certificate: This term is used in European Countries. This is an extension of the term given to pharmaceutical or plant protection patent.

Trade Name: The name under which a company conducts business, or by which its business, product/service are identified. It may or may not be registered as a trademark.

Trade Secret: A formula or process or device used in business that is not published or divulged which gives an advantage over competitors.

Trademark: A trademark can be a word, name, symbol, device or mark which is used to identify and distinguish the goods or services of one company from goods or services of another.

Traditional Knowledge: The Traditional Knowledge [TK] refers to the knowledge, innovations and practices of indigenous people and local communities.

B. BACKGROUND INFORMATION

1. The property refers to wealth or valuable things earned by a person. It is estimated in terms of land, house, garden, industries, animals, gold, silver, diamond, money, *etc.* accumulated by a person. The various valuable belongings of a person are broadly classified into two groups, *viz.*, (i) immovable property, and (ii) movable property.

2. Immovable Property refers to fixed types of properties which cannot move from one place to other. Such properties include land, buildings [house, flat, bungalow, farm house] and gardens. Such properties are protected by laws of land in each country.

3. Movable Property refers to the property which can be easily shifted from one place to other. Such property includes animals, farm machines,

furniture, fixtures, gold, silver, diamond, and money. Such properties are also protected by public laws.

4. There is third type of property what we call intellectual property. The product/process/idea which is outcome of the brain of a person and can be used on commercial scale for benefit of human kind is called intellectual property.

5. The Intellectual Property Rights are broadly divided into two groups, *viz.*, Primary Rights and Sui-generis Rights. Primary rights include,Copyrights, Patents, Trademarks, Industrial designs rights, utility models, geographical indication, trade secret, related rights, trade name and domain name.

6. The right of an author, artist, publisher, *etc.* to retain ownership of works and to produce or contract others to produce copies is called copyright. It is applicable to books, movies, music, paintings, photographs, sound recordings and software.

7. Patent refers to a document granting an inventor sole rights to an invention. Patent is applicable to new process, products, invention, device and innovations which have commercial or industrial applications.

8. A trademark or trade mark is a distinctive sign of some kind which is used by an individual, business organization or other legal entity to uniquely identify the source of its products and/or services to consumers, and to distinguish its products or services from those of other entities. A trademark includes a distinctive word, phrase, logo, domain name, graphic symbol, slogan, image or other device used to identify the source of a product from others.

9. An industrial design or simply a design is the ornamental or aesthetic aspect of an article produced by industry or handicraft. Industrial designs include any original shape, picture, or some combination applied to a useful article of manufacture.

10. A utility model is an exclusive right granted for an invention, which allows the right holder to prevent others from commercially using the protected invention, without his authorization, for a limited period of time. A utility model is similar to a patent. In fact, utility models are also referred to as "petty patents" or "innovation patents, minor patent and small patent.

11. A geographical indication is a sign used on goods that have a specific geographical origin and often possess qualities or a reputation that are due to that place of origin. In other words, a geographical indication is a name or sign used on certain products or which corresponds to a specific geographical location or origin (*e.g.*, a town, region, or country). It is used to identify agricultural, natural and manufactured goods.

12. Trade secret refers to a formula or process or device used in business that is not published or divulged which gives an advantage over competitors. A trade secret may consist of a formula for a chemical compound, a process of manufacturing, trading or preserving materials, a pattern for a machine or other device or a list of customers.

13. Related rights are rights which are similar to authors' right but which are not connected with the actual author of the work. Related rights or neighboring rights cover: performers' rights, the rights of phonogram producers, publishers' rights in typographical arrangements, the rights of broadcasting organizations in relation to their broadcasts, film producers, database creators, photographers and designers. Data base creators are also protected by related rights.

14. A trade name is the name under which a product is commercially known. The commercial name by which a chemical is known is called trade name. One chemical may have a variety of trade names depending on the manufacturers or distributors involved. Trade names cover all types of business, *viz.* sole proprietorship, partnership; corporate business, *etc.*

15. Domain refers to the unique name that identifies an Internet site. Domain names have two or more parts separated by dots. For example www.ask-edi.com or www.bcbsks.com.

16. Sui-generis is of its own kind/genus or unique in its characteristics. Such Intellectual Property Rights include (i) database Rights, (ii) mask work, (iii) moral rights, (iv) supplementary protection certificate (v), plant breeders rights, and (vii) traditional knowledge.

17. A 'database right' is an intellectual property right given to a computer database A mask work is a two or three-dimensional layout of an integrated circuit (IC or "chip")

18. Rights granted to the breeder of a new variety of plant are called plant breeders' rights.

19. The legal rights provided to farmers to save, use, sow, replant, exchange, share or sell his farm produce are called farmers' rights

20. Moral rights are a special extension to copyright that gives the originator of a copyrighted work rights over its use.

21. A supplementary protection certificate (SPC) is a sui generis, patent-like, intellectual property right that is available for medicinal and plant protection products.

22. The knowledge, innovations and practices of indigenous people and local communities are called traditional knowledge of indigenous property.

C. MULTIPLE CHOICE QUESTIONS

1. What type of property is gold?

(a) Immovable (b) Movable

(c) Intellectual (d) All of these

2. What type of property is a book?

(a) Patent (b) Trade mark

(c) Copy right (d) Petty right

3. Copy right is applicable to:

(a) Books (b) Movies

(c) Phonograms (d) All of these

4. Invention of new process or device comes under:

(a) Copy rights (b) Patent

(c) Trade mark (d) None of these

5. What type of property is a product formula?

(a) Copy right (b) Patent

(c) Trade secret (d) None of these

6. Sui-generis rights include:

(a) Data base rights (b) Moral rights

(c) Mask work (d) All of these

7. Sui-generis rights include:

(a) Plant breeders rights (b) Supplementary protection certificate

(c) Traditional Knowledge (d) All of these

8. Primary rights include:

(a) Copy rights (b) Patent rights

(c) Trade mark (d) All of these

9. Primary rights include:

(a) Trade secret (b) Industrial designs

(c) Utility models (d) All of these

10. Plant breeders rights come under:

(a) Primary rights (b) Sui-generis rights

(c) Both (a) and (b) (d) None of these

11. Farmers rights come under:

(a) Primary rights
(b) Sui-generis rights
(c) Both (a) and (b)
(d) None of these

12. What is the period of protection of copy rights?

(a) 25 years
(b) 40 years
(c) 60 years
(d) None of these

13. What is the period of protection of a patent?

(a) 20 years
(b) 10 years
(c) 5 years
(d) None of these

14. What is the period of protection of Trade mark?

(a) 5 years
(b) 10 years
(c) 15 years
(d) 20 years

15. What is the period of protection of Industrial design?

(a) 2 years
(b) 5 years
(c) 10 years
(d) 15 years

16. What is the period of protection of geographical indications?

(a) 5 years
(b) 10 years
(c) 15 years
(d) 20 years

17. What is the period of protection of Trade names?

(a) 2 years
(b) 3 years
(c) 5 years
(d) 10 years

18. What is the period of protection of domain name?

(a) 1 year
(b) 2 years
(c) 3 years
(d) None of these

19. What is the period of protection of registered mask work?

(a) 5 years
(b) 10 years
(c) 15 years
(d) 20 years

20. What is the period of protection of data base rights?

(a) 5 years
(b) 10 years
(c) 15 years
(d) None of these

D. FILL IN THE BLANKS

1. Utility models are also called ________.
2. Traditional knowledge is also known as ________.
3. Diamond comes under ________ property.
4. A new process comes under ________.
5. Books are covered by ________.
6. Crop varieties are covered by ________ rights.
7. Related rights are also known as ________ rights.
8. A trade name is also known as ________.
9. Related rights are very similar to ________.
10. A Formula belongs to ________.
11. Farmers are onsite conservators of ________.
12. PBR are covered by ________ rights.
13. Bansmati Rice is an example of ________.
14. Gardens come under ________ property.
15. Utility models come under primary ________.
16. The period of protection of data base rights is________.
17. The period of protection of copy rights is ________.
18. The period of protection of a patent rights is ________.
19. The period of protection of registered mask work rights is________.
20. The period of protection of trade secret is________.

E. MARK TRUE OR FALSE

1. Utility models are also called petty patent.
2. Traditional knowledge is also known as indigenous knowledge.
3. Diamond comes under movable property.
4. A new process comes under intellectual property.
5. Books are covered by copy rights.

6. Crop varieties are covered by plant breeders' rights.
7. Related rights are also known as neighboring rights.
8. A trade name is also known as business name.
9. Related rights are very similar to copy rights.
10. A Formula belongs to trade secret.
11. Farmers are onsite conservators of PGR.
12. PBR are covered by Sui-generis rights.
13. Bansmati Rice is an example of geographical indication.
14. Gardens come under immovable property.
15. Utility models come under primary intellectual property rights.
16. The period of protection of data base rights is 10 years.
17. The period of protection of copy rights is 60 years.
18. The period of protection of a patent rights is 20 years.
19. The period of protection of registered mask work rights is 10 years.
20. The period of protection of trade secret is indefinite.

F. ONE WORD ANSWER

1. The wealth or valuable things earned by a person.
2. Fixed types of properties which cannot move from one place to other.
3. The property which can be easily shifted from one place to other.
4. The product/process/idea which is outcome of the brain of a person and can be used on commercial scale for benefit of human kind.
5. The legal rights provided to an inventor to derive economic benefits from his invention/innovation.
6. The exclusive rights to reproduce, sell and distribute a work, prepare derivative works and display the work publicly.
7. A document granting an inventor sole rights to an invention.
8. A symbol, device or mark which is used to identify and distinguish the goods or services of one company from goods or services of another.

9. A formula or process or device used in business that is not published or divulged which gives an advantage over competitors.
10. The name under which a company conducts business, or by which its business, product/service are identified.
11. The names and words that companies designate for their registered Internet Web site addresses.
12. Any original shape, picture, or some combination applied to a useful article of manufacture.
13. A sign used on goods that have a specific geographical origin and often possess qualities or a reputation that are due to that place of origin.
14. The legal privileges that only a copyright owner has with respect to their copyrighted work.
15. Things of their own kind or things with unique characteristics.
16. An intellectual property right given to a computer database.
17. A two or three-dimensional layout of an integrated circuit.
18. Special extension to copyright that gives the originator of a copyrighted work rights over its use.
19. An extension of the term given to pharmaceutical or plant protection patent.
20. The knowledge, innovations and practices of indigenous people and local communities.
21. Intellectual property rights granted to the breeder of a new variety of plant.
22. The legal rights provided to farmers to save, use, sow, replant, exchange, share or sell his farm produce including seed of a variety protected under Plant

ANSWERS

MULTIPLE CHOICE QUESTIONS

Sl. No.	*Answer*	*Sl. No.*	*Answer*
1	(b) Movable	2	(c) Copy right
3	(d) All of these	4	(c) Trade mark
5	(c) Trade secret	6	(d) All of these
7	(d) All of these	8	(d) All of these
9	(d) All of these	10	(b) Sui-generis rights
11	(b) Sui-generis rights	12	(c) 60 years
13	(a) 20 years	14	(c) 60 years
15	(b) 5 years	16	(b) 10 years
17	(b) 3 years	18	(a) 1 years
19	(b) 10 years	20	(c) 15 years

FILL IN THE BLANKS

Sl. No.	*Answer*	*Sl. No.*	*Answer*
1	Petty patent	2	Indigenous knowledge
3	Movable	4	Patent
5	Copy right	6	Sui-generis
7	Neighboring	8	Trade Name
9	Copy right	10	Trade Secret
11	Plant Genetic Resources	12	Sui-generis rights
13	Geographical indication	14	Immovable
15	IPR	16	10 years
17	60 years	18	20 years
19	10 years	20	Indefinite

TRUE OR FALSE

1 to 20 all are true

APPROPRIATE TERM/ONE WORD ANSWER

Sl. No.	*Answer*	*Sl. No.*	*Answer*
1	Property	2	Immovable Property
3	Movable Property	4	Intellectual Property
5	Intellectual Property Rights	6	Copyrights
7	Patent	8	Trade Mark
9	Trade Secret	10	Trade Name
11	Domain Name	12	Industrial Design
13	Geographical Indication	`14	Exclusive Rights
15	Sui Generis Rights	16	Database Rights
17	Mask Work	18	Moral Rights
19	Supplementary Protection Certificate	20	Traditional Knowledge
21	Plant Breeders' Rights	22	Farmers' Rights

*****

Chapter 15

Modern Tools for Line and Cultivar Development [Haploids, Somtic hybridization, Tissue Culture and Biotechnology]

A: DEFINITIONS

[a] Haploidy

Allohaploid: Haploid that develop from allopolyploid species.

Aneuhaploid: Haploid that develop from an aneuploid species.

Autohaploid: Haploid that develops from autoployploid species.

Euhaploid: Haploid which develop from a euploid species.

Haploid: An individual with half of somatic chromosome number.

Monohaploid: Haploid which develops from a normal diploid species.

Monoploids: Individuals with basic chromosome number of a species. Such condition is called monoploidy.

Polyhaploid: Haploid that develop from a polyploid species.

[b] Somatic Hybridization

Asymmetrical somatic hybrids: Somatic hybrids that contain complete somatic complement of one species and only part of the somatic complements of other species.

Cybrids: Somatic hybrids with normal protoplast of one species and nucleusless protoplast of other species.

Cytoplast: A protoplast either without nucleus or with inactive nucleus.

Heterokaryons: Hybrid cell involving protoplasts of two different species.

Homokaryons: Hybrid cell involving protoplasts of the same species.

Inter-generic somatic hybrids: Hybrids obtained by protoplast fusion between two different genera of the same family.

Inter-specific somatic hybrids: Hybrids obtained by protoplast fusion between two different species of the same genus.

Inter-tribal somatic hybrids: Hybrids obtained through protoplast fusion between plants of two different families.

Protoplast: Naked cells or cells without cell wall.

Somatic hybridization: Crossing of crop plants through fusion of somatic cells.

Somatic Hybrids: Hybrids obtained by somatic hybridization.

Symmetrical somatic hybrids: Somatic hybrids that contain all chromosomes of both the species involved in the fusion of protoplast.

[c] Tissue Culture

Callus: An unorganized mass of differentiated plant cells.

Clonal propagation: asexual multiplication of plants from a single individual or explant.

Culture: A plant growing *in vitro* in a sterile environment.

Doubled Haploid: Plants derived from a haploid after doubling chromosome number by colchicine treatment to form homozygous diploids.

Embryo Rescue: In vitro culture technique used to rescue inherently weak, immature or hybrid embryos to prevent degeneration.

***Ex vitro*:** Organisms removed from tissue culture and transplanted; generally plants to soil or potting mixture.

Explant: An excised piece or part of a plant used to initiate a tissue culture.

Micro-propagation: Multiplication of plants from vegetative parts by using tissue culture nutrient medium.

Protoplast: Cells without cell wall, usually removed by digestion with enzymes.

Somaclonal Variation: Genetic variation among progeny of plants regenerated from somatic cells cultured *in vitro*.

Somatic embryos: non-zygotic bipolar embryo-like structures obtained from somatic cells.

Tissue Culture: Growth of tissues or cells in an artificial medium

Totipotency: Capacity of plant cells to regenerate whole plants when cultured on media.

Transgenic: Plants that have a piece of foreign DNA

[d] Transgenic Technology

Agrobacterium tumifaciens: A soil borne bacterium which is widely used for development of transgenic plants in different crop plants. It is a biological method of foreign gene transfer into host cells.

Biotechnology: Cellular and bio-molecular processes to develop technologies and products that can be commercialized.

Gene Cloning: The process of making several identical copies of a gene in genetic engineering.

Genetic Engineering: Isolation, introduction and expression of foreign DNA in Plants and animals.

Genetic Transformation: The process of transfer, integration and expression of transgene (foreign gene) in an organism.

Micro-Injection: A mechanical method of gene transfer in which foreign DNA is delivered into the host cells by microscopic needles also known as micro-injection.

Particle Bombardment: A method of foreign gene transfer which involves high velocity metal particles (usually gold or tungsten) or metal particles mediated method of genetic transformation. It is also known as biolistic micro projectile bombardment, micro ballistics and particle acceleration.

Plant Biotechnology: Combined study of plant tissue culture and genetic engineering or recombinant DNA technology.

Plasmid: An extra-chromosomal genetic element which is found in the bacteral cell, can replicate independently of chromosomal DNA and is not essential for normal growth and development of bacterium.

Transformants: Cells or tissues of an organism which have udergone genetic transformation; resultants of genetic transformation.

Transgenes: Foreign genes or modified genes of the same species which are used for the development of transgenic individuals. Such gene may be from the same species (in modified form), related wild species, unrelated species and microbes (bacteria, fungi and viruses).

Transgenic Breeding: Genetic improvement of crop plants, domestic animals and useful micro-organisms through biotechnology.

Transgenic: Genetically engineered organisms or organisms developed by the techniques of genetic engineering. It may be a plant, an animal or micro-organism.

B: BACKGROUND INFORMATION

[a] Haploidy

1. Important alternative strategies include, (i) haploids, (ii) tissue culture, (iii) somatic hybridization and (iv) transgenic technology.

2. An individual with gametic chromosome number is known as haploid and such condition is referred to as haploidy.

3. There is difference between haploids and monoploids. Monoploids have the basic chromosome number (x) of a species. Haploids have gametic chromosome number (n).

4. In a true diploid species, both monoploid and haploid chromosome number is the same ($n = x$). Thus a monoploid can be a haploid but all haploids cannot be monoploids.

5. In crop plants, haploids may originate in several ways, *viz.* (1) by parthenogenesis, (2) from interspecific crosses, (3) as a member of twin embryos, (4) by semigamy, (5) by alien cytoplasm, and (6) by anther-culture.

6. Haploids have several applications in plant breeding such as (1) development of pure lines, (2) disease resistance, (3) development of inbred lines, and (4) indirect uses.

[b] Tissue Culture

7. Plant tissue culture refers to growing of living plant cells, tissues or organ on a suitable nutrient medium outside the tissues of parent plant.

8. Plant tissue culture has a number of commercial applications such as micro-propagation, somatic hybridization, embryo rescue, production of haploids, development of transgenic plants, selection of resistant plants, virus free plants, production of synthetic seeds, conservation of elite material and production of secondary metabolites.

[c] Somatic Hybridization

9. Somatic hybridization refers to crossing of crop plants through fusion of somatic cells.

10. The fusion of somatic cells takes place through protoplasts. Naked cells or cells without cell wall are called protoplasts.

11. There are several potential applications of somatic hybridization in plant breeding such as gene transfer across species, development of somatic hybrids, conservation of heterosis, development of transgenic plants, production of fertile plants, modification of cytoplasmic genes, enhancement of photosynthetic efficiency, production of allotetraploid plants and hybrids in asexually propagated species.

[d] Transgenic Technology

12. The technique or process of developing transgenic individuals is referred to as transgenic technology. It is also known as gene technology,

transformation technology, genetic engineering and recombinant DNA technology.

13. The crop plants which are developed through the application of plant biotechnology or plant genetic engineering are called transgenic plants or genetically modified plants or genetically engineered plants.

14. In broad sense transgenic plants can be called as genetically modified plants. But in strict sense, they are genetically engineered plants. However, the term GM crops or GM plants or GM organisms has been very popular and is widely used the world over.

15. The GM plants have been developed to solve specific problems which could not be solved easily through conventional plant breeding methods.

16. Important applications of transgenic technology include improvement in (i) biotic resistance, (ii) abiotic resistance, (iii) herbicide resistance, (iv) quality of food products and (v) development of novel traits such as golden rice and male sterility.

C: MULTIPLE CHOICE QUESTIONS

1. How many basic sets of chromosomes are there in a genome?

(a) One (b) Two
(c) Three (d) Four

2. How are monoploids represented?

(a) x (b) 2x
(c) n (d) 2n

3. From which of the following species dihaploid develops?

(a) Diploid (b) Triploid
(c) Tetraploid (d) All of these

4. A cell without cell wall is called:

(a) Proplastid (b) Plastid
(c) Protoplast (d) Cytoplast

5. A naked cell is called:

(a) Proplastid (b) Plastid
(c) Protoplast (d) Cytoplast

6. Organ culture refers to:

(a) Embryo culture (b) Anther culture
(c) Ovule culture (d) All of these

7. Tissue culture refers to:

(a) Anther culture (b) Protoplast culture

(c) Ovule culture (d) All of these

8. Callus refers to a mass of:

(a) Cells (b) Unorganized cells

(c) Organized cells (d) All of these

9. Somatic hybridization may be:

(a) Interspecific (b) Intergeneric

(c) Inter-tribal (d) All of these

10. Transgenes can be utilized from:

(a) Bacteria (b) Fungi

(c) Viruses (d) All of these

D: FILL IN THE BLANKS

1. In a genome each type of chromosome is represented ________.
2. ________ have single copy of the genome.
3. ________ is represented by x.
4. ________ is represented by n.
5. All hapoids are not ________.
6. All ________ are haploids.
7. Dihploid develops from a ________ species.
8. Diploids have ________ basic sets of a genome.
9. Doubled ________are developed by doubling chromosome number of a haploid.
10. Somatic hybridization involves ________fusion.
11. ________involve protoplasts of the species.
12. ________ involve protoplasts of two species.
13. ________ is a protoplast either without nucleus or with inactive nucleus.
14. ________ are cells without cell wall.
15. Cytoplasmic hybrids are also known as ________.
16. ________is an unorganized mass of differentiated plant cells.

17. ________is an excised piece or part of a plant used to initiate a tissue culture.

18. ________ is the capacity of plant cells to regenerate whole plants when cultured on media.

19. ________ refers to growing of tissues or cells in an artificial medium.

20. ________ are organisms developed by the techniques of genetic engineering.

E: MARK TRUE OR FALSE

1. In a genome each type of chromosome is represented only once.
2. Monoploids have single copy of the genome.
3. Monoploid is represented by x.
4. Haploid is represented by n.
5. All hapoids are not monoploids.
6. All monoploids are haploids.
7. Dihaploid develops from a tetraploid species.
8. Diploids have two basic sets of a genome.
9. Doubled haploids are developed by doubling chromosome number of a haploid.
10. Somatic hybridization involves fusion of protoplast of two different species.
11. Homokaryons involve protoplasts of the same species.
12. Heterokaryons involve protoplasts of the same species.
13. Cytoplast is a protoplast either without nucleus or with inactive nucleus.
14. Protoplasts are cells without cell wall.
15. Cytoplasmic hybrids are also known as cybrids.
16. Callus is an unorganized mass of differentiated plant cells.
17. Explant is an excised piece or part of a plant used to initiate a tissue culture.
18. Totipotency is the capacity of plant cells to regenerate whole plants when cultured on media.

19. Tissue culture refers to growth of tissues or cells in an artificial medium
20. Transgenics are organisms developed by the techniques of genetic engineering.

F: SORT OR ONE WORD ANSWER

1. An individual with half of somatic chromosome number.
2. Individuals with basic chromosome number of a species. Such condition is called monoploidy.
3. Haploid which develop from a euploid species.
4. Haploid which develops from a normal diploid species.
5. Haploid that develop from a polyploid species.
6. Haploid that develop from allopolyploid species.
7. Haploid that develops from autoploypoid species.
8. Haploid that develop from an aneuploid species.
9. Crossing of crop plants through fusion of somatic cells.
10. Hybrids obtained by somatic hybridization.
11. Hybrids obtained by protoplast fusion between two different species of the same genus.
12. Hybrids obtained by protoplast fusion between two different genera of the same family.
13. Hybrids obtained through protoplast fusion between plants of two different families.
14. Naked cells or cells without cell wall.
15. Somatic hybrids that contain all chromosomes of both the species involved in the fusion of protoplast.
16. Somatic hybrids that contain complete somatic complement of one species and only part of the somatic complements of other species.
17. Somatic hybrids with normal protoplast of one species and nucleusless protoplast of other species.
18. Hybrid cell involving protoplasts of the same species.
19. Hybrid cell involving protoplasts of two different species.

20. A protoplast either without nucleus or with inactive nucleus.
21. An unorganized mass of differentiated plant cells.
22. Asexual multiplication of plants from a single individual or explant.
23. A plant growing *in vitro* in a sterile environment.
24. Plants derived from a haploid after doubling chromosome number by colchicine treatment to form homozygous diploids.
25. In vitro culture technique used to rescue inherently weak, immature or hybrid embryos to prevent degeneration.
26. An excised piece or part of a plant used to initiate a tissue culture.
27. Multiplication of plants from vegetative parts by using tissue culture nutrient medium.
28. Cells without cell wall, usually removed by digestion with enzymes.
29. Non-zygotic bipolar embryo-like structures obtained from somatic cells.
30. Genetic variation among progeny of plants regenerated from somatic cells cultured *in vitro*.
31. Capacity of plant cells to regenerate whole plants when cultured on media.
32. Plants that have a piece of foreign DNA.
33. Growth of tissues or cells in an artificial medium.
34. Cellular and bio-molecular processes to develop technologies and products that can be commercialized.
35. Foreign genes or modified genes of the same species which are used for the development of transgenic individuals.
36. Genetically engineered organisms or organisms developed by the techniques of genetic engineering. It may be a plant, an animal or micro-organism.
37. The process of transfer, integration and expression of transgene (foreign gene) in an organism.
38. Cells or tissues of an organism which have undergone genetic transformation; resultants of genetic transformation.
39. An extra-chromosomal genetic element which is found in the bacterial cell, can replicate independently of chromosomal DNA and is not essential for normal growth and development of bacterium.

40. Combined study of plant tissue culture and genetic engineering or recombinant DNA technology.

41. Isolation, introduction and expression of foreign DNA in Plants and animals.

42. The process of making several identical copies of a gene in genetic engineering.

43. A metal particles mediated method of genetic transformation. It is also known as biolistic micro projectile bombardment, micro ballistics and particle acceleration.

44. A soil borne bacterium which is widely used for development of transgenic plants in different crop plants. It is a biological method of foreign gene transfer into host cells.

45. A mechanical method of gene transfer in which foreign DNA is delivered into the host cells by microscopic needles also known as micro-injection.

46. Genetic improvement of crop plants, domestic animals and useful micro-organisms through biotechnology.

ANSWERS

MULTIPLE CHOICE QUESTIONS

Sl. No.	*Answer*	*Sl. No.*	*Answer*
1	(a) One	2	(a) x
3	(c) Tetraploid	4	(c) Protoplast
5	(c) Protoplast	6	(d) All of these
7	(d) All of these	8	(b) Unorganized cells
9	(d) All of these	10	(d) All of these

FILL IN THE BLANKS

Sl. No.	*Answer*	*Sl. No.*	*Answer*
1	Once	2	Monoploid
3	Monoploid	4	Haploid
5	Monoploid	6	Haploids
7	Tetraploid	8	Two
9	Haploids	10	Protoplast
11	Homokayrones	12	Heterokaryones
13	Cytoplast	14	Protoplast
15	Cybrids	16	Callus
17	Explant	18	Totipotency
19	Tissue culture	20	Transgenics

TRUE OR FALSE

1 to 20 all are true.

APPROPRIATE TERM/ONE WORD ANSWER

Sl. No.	*Answer*	*Sl. No.*	*Answer*
1	Haploid	2	Monoploids
3	Euhaploid	4	Monohaploid
5	Polyhaploid	6	Allohaploid
7	Autohaploid	8	Aneuhaploid

Sl. No.	*Answer*	*Sl. No.*	*Answer*
9	Somatic hybridization	10	Somatic Hybrids
11	Inter-specific somatic hybrids	12	Inter-generic somatic hybrids
13	Inter-tribal somatic hybrids	14	Protoplast
15	Symmetrical somatic hybrids	16	Asymmetrical somatic hybrids
17	Cybrids	18	Homokaryons
19	Heterokaryons	20	Cytoplast
21	Callus	22	Clonal propagation
23	Culture	24	Doubled Haploid
25	Embryo Rescue	26	Explant
27	Micro-propagation	28	Protoplast
29	Somatic embryos	30	Somaclonal Variation
31	Totipotency	32	Transgenic
33	Tissue Culture	34	Biotechnology
35	Transgenes	36	Transgenic
37	Genetic Transformation	38	Transformants
39	Plasmid	40	Plant Biotechnology
41	Genetic Engineering	42	Gene Cloning
43	Particle Bombardment	44	Agrobacterium tumifaciens
45	Micro-Injection	46	Transgenic Breeding

*****

Appendices

Appendix 1
List of Agricultural Universities in India

I. AGRICULTURAL UNIVERSITIES

Sl.No.	Name of University	Location	State
A	STATE UNIVERSITIES		
1	Acharya NG Ranga Agricultural University.	Hyderabad	Telangana
2	Assam Agricultural University.	Jorhat	Assam
3	Bihar Agricultural University.	Sabour	Bihar
4	Birsa Agricultural University	Ranchi	Jharkhand
5	Indira Gandhi Krishi Vishwavidyalaya.	Raipur	Chhattisgarh
6	Anand Agricultural University.	Anand	Gujarat
7	Junagadh Agricultural University.	Junagarh	Gujarat
8	Navsari Agricultural University.	Navsari	Gujarat
9	Sardarkrushinagar-Dantiwada Agricultural University.	Bansakantha	Gujarat
10	Chaudhary Charan Singh Haryana Agricultural University.	Hisar	Haryana
11	CSK Himachal Pradesh Krishi Vishvavidyalaya	Palampur	Himachal Pradesh
12	Sher-e-Kashmir Univ of Agricultural Sciences and Technology.	Jammu	Jammu and Kashmir
13	Sher-e-Kashmir Univ of Agricultural Sciences and Technology of Kashmir.	Srinagar	Jammu and Kashmir
14	University of Agricultural Sciences.	Bangalore	Karnataka
15	University of Agricultural Sciences.	Dharwad	Karnataka
16	University of Agricultural Sciences.	Raichur	Karnataka
17	Kerala Agricultural University.	Vallanikara	Kerala
18	Jawaharlal Nehru Krishi Viswavidyalaya.	Jabalpur	Madhya Pradesh
19	Rajmata Vijayaraje Scindia Krishi Viswavidyalaya.	Gwalior	Madhya Pradesh
20	Dr. Balasaheb Sawant Konkan Krishi Vidyapeeth.	Ratnagiri	Maharashtra
21	Dr. Panjabrao Deshmukh Krishi Vidyapeeth.	Akola	Maharashtra
22	Mahatma Phule Krishi Vidyapeeth.	Rahuri	Maharashtra
23	Marathwada Agricultural University.	Parbhani	Maharashtra
24	Odisha Univ. of Agriculture and Technology.	Bhubaneswar	Odisha
25	Punjab Agricultural University.	Ludhiana	Punjab
26	Maharana Pratap Univ. of Agriculture and Technology.	Udaipur	Rajasthan
27	Rajasthan Agricultural University.	Bikaner	Rajasthan
28	Agriculture University.	Kota	Rajasthan
29	Sri Karan Narendra Agriculture University.	Jobner	Rajasthan
30	Agriculture University.	Jodhpur	Rajasthan
31	Tamil Nadu Agricultural University.	Coimbatore	Tamil Nadu

Sl.No.	Name of University	Location	State
32	Chandra Shekar Azad University of Agriculture and Technology.	Kanpur	Uttar Pradesh
33	Narendra Deva University of Agriculture and Technology.	Faizabad	Uttar Pradesh
34	Sardar Ballabh Bhai Patel Univ. of Agriculture and Technology.	Meerut	Uttar Pradesh
35	Manyavar Kashiram University of Agriculture and Technology.	Banda	Uttar Pradesh
36	Govind Ballabh Pant University of Agriculture and Technology.	Pantnagar	Uttrakhand
37	Bidhan Chandra Krishi Viswavidyalaya.	Kalyani	West Bengal
38	Uttar Banga Krishi Viswavidyalaya.	Cooch Bihar	West Bengal
B	CENTRAL UNIVERSITIES		
1	Central Agricultural University.	Imphal	Manipur
2	Rani Laxmibai Central Agricultural University.	Jhansi	Uttar Pradesh
3	Rajendra Central Agricultural University.	Pusa	Bihar
C	DEEMED UNIVERSITIES		
1	Indian Agricultural Research Institute.	Pusa	New Delhi
2	National Dairy Research Institute.	Karnal	Haryana
3	Indian Veterinary Research Institute.	Bareilly	Uttar Pradesh
4	Central Institute for Fisheries Education.	Mumbai	Maharashtra
5	Allahabad Agricultural Institute.	Allahabad	Uttar Pradesh

II. ANIMAL SCIENCE UNIVERSITIES

Sl.No.	Name of University	Location	State
1	Sri Venkateswara Veterinary University.	Tirupati	Andhra Pradesh
2	Karnataka Veterinary, Animal and Fisheries Sciences University.	Bidar	Karnataka
3	Maharashtra Animal Science and Fishery University.	Nagpur	Maharashtra
4	Guru Angad Dev Veterinary and Animal Science University.	Ludhiana	Punjab
5	Tamil Nadu Veterinary and Animal Science University.	Chennai	Tamil Nadu
6	Tamil Nadu Fisheries University.	Nagapattinam	Tamil Nadu
7	UP Pandit Deen Dayal Upadhaya Pashu Chikitsa Vigyan Vishwa Vidhyalaya evam Go Anusandhan Sansthan, Mathura.	Mathura	Uttar Pradesh
8	West Bengal University of Animal and Fishery Sciences.	Kolkata	West Bengal

9	Rajasthan University of Veterinary and Animal Sciences.	Bikaner	Rajasthan
10	Kerala University of Fisheries and Ocean Studies.	Kochi	Kerala
11	Kerala Veterinary and Animal Sciences University.	Wayanand	Kerala

III. HORTICULTURAL UNIVERSITIES

Sl.No.	Name of University	Location	State
1	Andhra Pradesh Horticultural University.	Tadepalligudem	Andhra Pradesh
2	University of Horticultural Sciences.	Bagalkot	Karnataka
3	Dr. Yashwant Singh Parmar Univ of Horticulture and Forestry.	Solan	Himachal Pradesh
4	Dr. YSR Horticultural University.	Veketaramanna-gudem	Andhra Pradesh
5	Uttarakhand University of Horticulture and Forestry.	Bharsar	Uttarakhand

Appendix 2
List of ICAR Research Institutes in India

Sl.No.	Name of Research Institute	Location	State
A	CROP SCIENCES		
1	ICAR-Central Institute for Cotton Research	Nagpur	Maharashtra
2	ICAR- Central Institute for Jute and Allied Fibres	Kolkata	West Bengal
3	ICAR-National Rice Research Institute	Cuttack	Odisha
4	Central Tobacco Research Institute	Rajamundry	Andhra Pradesh
5	ICAR- Directorate of Groundnut Research	Junagarh	Gujarat
6	ICAR-Indian Institute of Oilseeds Research	Hyderabad	Telangana
7	ICAR- Directorate of Rapeseed-Mustard Research	Bharatpur	Rajasthan
8	ICAR-Indian Institute of Rice Research	Hyderabad	Talangana
9	ICAR-Indian Institute of Seed Science	Mau	Uttar Pradesh
10	ICAR-Indian Institute of Millets Research	Hyderabad	Talangana
11	ICAR-Indian Institute of Soybean Research	Indore	Madhya Pradesh
12	ICAR-Indian Agriculture Research Institute	Pusa	New Delhi
13	ICAR-Indian Agriculture Research Institute	Hazaribagh	Jharkhand
14	ICAR-Indian Grassland and Fodder Research Institute	Jhansi	Uttar Pradesh
15	ICAR-Indian Institute of Maize Research	Ludhiana	Punjab
16	ICAR-Indian Institute of Agricultural Biotechnology	Ranchi	Bihar
17	ICAR-Indian Institute of Pulses Research	Kanpur	Uttar Pradesh
18	ICAR-Indian Institute of Sugarcane Research	Lucknow	Uttar Pradesh
19	ICAR-Indian Institute of Wheat and Barley Research	Karnal	Haryana
20	ICAR-National Bureau of Agricultural Insect Resources	Bangaluru	Karnataka
21	ICAR-National Bureau of Agriculturally Important Micro-organism	Mau	Uttar Pradesh
22	ICAR-National Bureau of Plant Genetic Resources	Pusa	New Delhi
23	ICAR-National Centre for Integrated Pest Management	Pusa	New Delhi
24	ICAR- National Institute of Biotic Stress Management	Raipur	Chhattisgarh
25	ICAR-National Centre on Plant Biotechnology	Pusa	New Delhi
26	ICAR-Sugarcane Breeding Institute	Coimbatore	Tamil Nadu
27	ICAR-Vivekanand Parvatiya Krishi Anusandhan Sansthan	Almora	Uttrakhand
B	HORTICULTURAL SCIENCES		
28	ICAR-Central Institute for Arid Horticulture	Bikaner	Rajasthan
29	ICAR-Central Institute for Subtropical Horticulture	Lucknow	Uttar Pradesh
30	ICAR-Central Institute for Temperate Horticulture	Srinagar	Jammu and Kashmir

Sl.No.	Name of Research Institute	Location	State
31	ICAR- Central Island Agricultural Research Institute	Portblair	Andman and Nicobar
32	ICAR-Central Plantation Crops Research Institute	Kasargod	Kerala
33	ICAR-Central Potato Research Institute	Shimla	Himachal Pradesh
34	ICAR-Central Tuber Crops Research Institute	Trivendrum	Kerala
35	ICAR-Directorate of Cashew Research	Puttur	Karnataka
36	ICAR-Directorate of Floriculture Research	Pusa	New Delhi
37	ICAR-Directorate of Medicinal And Aromatic Plants Research	Anand	Gujarat
38	ICAR-Directorate of Mashroom Research	Solan	Himachal Pradesh
39	ICAR-Directorate of Onion and Garlic Research	Pune	Maharashtra
40	ICAR-Indian Institute of Horticulture Research	Bangalore	Karnataka
41	ICAR-Indian Institute of Oil Palm Research	Pedavegi	Andhra Pradesh
42	ICAR-Indian Institute of Spices Research	Calicut	Kerala
43	ICAR-Indian Institute of Vegetable Research	Varanasi	Uttar Pradesh
44	ICAR-National Research Centre for Banana	Tiruchirapalli	Kerala
45	ICAR-Central Citrus Research Institute	Nagpur	Maharashtra
46	ICAR-National Research Centre for Grapes	Pune	Maharashtra
47	ICAR-National Research Centre for Litchi	Muzaffarpur	Bihar
48	ICAR-National Research Centre for Orchids	Gangtok	Sikkim
49	ICAR-National Research Centre on Pomegranate	Solapur	Maharashtra
C	NATURAL RESOURCE MANAGEMENT		
50	ICAR-National Research Centre for Seed Spices	Ajmer	Rajasthan
51	ICAR-Central Agro-forestry Research Institute	Jhansi	Uttar Pradesh
52	ICAR-Central Arid Zone Research Institute	Jodhpur	Rajasthan
53	ICAR-Central Research Institute for Dry land Agriculture	Hyderabad	Andhra Pradesh
54	ICAR-Central Soil and Water Conservation	Dehradun	Uttarakhand
55	ICAR-Central Soil Salinity Research Institute	Karnal	Haryana
56	Indian Institute of Water Management	Bhubaneswar	Odisha
57	Directorate of Weed Research	Jabalpur	Madhya Pradesh
58	ICAR Research Complex for NEH Region	Shillong	Meghalaya
59	ICAR Research Complex for Eastern Region	Patna	Bihar
60	ICAR-Indian Institute of Soil Science	Bhopal	Madhya Pradesh
61	ICAR-Indian Institute of Farming System Research	Meerut	Uttar Pradesh
62	ICAR-Indian Institute of Soil Science	Bhopal	Madhya Pradesh
63	ICAR-National Bureau of Soil Survey and Land Use Planning	Nagpur	Maharashtra
64	ICAR-National Institute of Abiotic Research management	Baramati	Maharashtra

Sl.No.	Name of Research Institute	Location	State
65	ICAR- National Organic Farming Research	Gangtok	Sikkim
66	ICAR_ National Research Centre for Integrated Farming	Motihari	Bihar
C	AGRICULTURAL ENGINEERING		
67	ICAR-Central Institute for Research on Cotton Technology	Mumbai	Maharashtra
68	ICAR-Central Institute of Agricultural Engineering	Bhopal	Madhya Pradesh
69	ICAR-Central Institute Post Harvest Engineering and Technology	Ludhiana	Punjab
70	ICAR-Indian Institute of Natural Resigns and Gums	Ranchi	Jharkhand
71	ICAR-National Institute of Research on Jute and Allied Fibres Technology	Kolkata	West Bengal
D	AGRICULTURAL EDUCATION		
72	ICAR-Central Institute for Women in Agriculture	Bhubaneswar	Odisha
73	ICAR-Indian Agricultural Statistics Research Institute	Pusa Campus	New Delhi
74	ICAR-National Academy of Agricultural Research Management	Hyderabad	Telangana
75	ICAR-National Institute of Agricultural Economics and Policy Research	Pusa Campus	New Delhi
E	AGRICULTURAL EXTENSION		
76	ICAR-Directorate of Knowledge Management in Agriculture	Pusa Campus	New Delhi
F	ANIMAL RESEARCH INSTITUTES		
77	ICAR-Central Avian Research Institute	Bareilly	Uttar Pradesh
78	ICAR-Central Institute for Research on Buffaloes	Hisar	Haryana
79	ICAR-Central Institute for Research on Cattle	Meerut	Uttar Pradesh
80	ICAR-Central Institute for Research on Goats	Mathura	Uttar Pradesh
81	ICAR-Central Sheep and Wool Research Institute	Avikanagar	Rajasthan
82	ICAR-Directorate of Poultry Research	Hyderabad	Telangana
83	ICAR-Indian Veterinary Research Institute	Bareilly	Uttar Pradesh
84	ICAR-National Bureau of Animal Genetic Resources	Karnal	Haryana
85	ICAR-National Dairy Research Institute	Karnal	Haryana
86	ICAR-National Institute of Animal Nutrition and Physiology	Karnal	Haryana
87	ICAR-National Institute of High Security Animal Diseases	Bhopal	Madhya Pradesh
88	ICAR-National Institute of Veterinary Epidemiology and Disease Informatics	Bangalore	Karnataka
89	ICAR-National Research Centre for Equines	Hisar	Haryana
90	ICAR-National Research Centre for Camel	Bikaner	Rajasthan
91	ICAR-National Research Centre on Meat	Hyderabad	Telangana
92	ICAR-National Research Centre on Mithun	Jharnapani	Nagaland

Sl.No.	Name of Research Institute	Location	State
93	ICAR-National Research Centre on Pig	Guwahati	Assam
94	ICAR-National Research Centre on yak	Dirang	Arunachal Pradesh
95	ICAR-Project Directorate on Foot and Mouth Disease	Nainital	Uttarakhand
H	FISHERIES INSTITUTES		
96	ICAR-Central Island Fisheries Research Institute	Barrackpore	West Bengal
97	ICAR-Central Institute of Brackish water Aquaculture	Chennai	Tamil Nadu
98	ICAR-Central Institute Fisheries Education	Mumbai	Maharashtra
99	ICAR-Central Institute Fisheries Technology	Cochin	Kerala
100	ICAR-Central Institute of Fresh Water Aquaculture	Bhubaneswar	Odisha
101	ICAR-Central Marine Fisheries Research Institute	Kochi	Kerala
102	ICAR-Directorate of Cold Water Fisheries Research	Bhimtal	Uttarakhand
103	ICAR-National Bureau of Fish Genetic Resources	Lucknow	Uttar Pradesh

*****

Appendix 3
List of International Agricultural Research Institutes/Centres

Sl.No.	*Name of Research Institute/Centre*	*Founded in*	*Location*	*Country*
1.	International Rice Research Institute [IRRI]	1960	Manila	Philippines
2.	International Wheat and Maize Improvement Centre [CIMMYT]	1963	Mexico City	Mexico
3.	International Centre for Tropical Agriculture [CIAT]	1967	Cali	Colombia
4.	International Institute for tropical Agriculture [IITA]	1967	Ibadan	Nigeria
5.	International Potato Centre [CIP]	1971	Lima	Peru
6.	International Crop Research Institute for Semiarid Tropics [ICRISAT]	1972	Hyderabad	India
7.	International Centre for Agriculture Research in Drylans Areas [ICARDA]	1977	Aleppo	Syria
8.	West Africa Rice Development Association [WARDA]	1970	Bouake	Cote divoire
9.	International Plant Genetic Resources Institute [IPGRI]	1974	Rome	Italy
10.	Centre for International Forestry Research [CIFOR]	1992	Jakarta	Indonesia
11.	International Centre for Research in Agroforestry [ICRAF]	1991	Nairobi	Kenya
12.	International Centre for Living Aquatic Resources Management [ICLARM]	1977	Makati City	Philippines
13.	International Food Policy Research Institute [IFPRI]	1975	Washington DC	USA
14.	International Irrigation Management Institute [IIMI]	1984	Colombo	Sri Lanka
15.	International Livestock Research Institute [ILRI]	1995	Nairobi	Kenya
16.	International service for National Agricultural research [ISl.No.AR]	1979	Hague	Netherlands
17.	Asian Vegetable Research and Development Centre [AVRDC]	1971	Taiwan City	Taiwan [China]
18.	International Centre for Genetic Engineering and Biotechnology [ICGEB]	1987	Trieste	Italy

*****

Appendix 4
List of Indian Seed Testing Laboratories

Sl.No.	Name of Laboratory	Location/Address	State
1	State Seed Testing Laboratory	Ranvir Singh Pura, Jammu	Jammu and Kashmir
2	Seed Testing Laboratory	Lalmandi, Srinagar	Jammu and Kashmir
3	State Seed Testing Laboratory	Palampur	Himachal Pradesh
4	Seed Testing Laboratory	Solan	Himachal Pradesh
5	Seed Testing Laboratory	PAU, Ludhiana	Punjab
6	State Seed Testing Laboratory	HAU, Hisar	Haryana
7	Seed Testing Laboratory	Karnal	Haryana
8	Quality Control Laboratory	Umri, Kurukshetra	Haryana
9	Seed Testing Laboratory	Durgapura, Jaipur	Rajasthan
10	State Seed Testing Laboratory	Agri Uni., Kanpur	Uttar Pradesh
11	Seed Testing Laboratory	Agri. Uni., Pantnagar	Uttar Khand
12	Seed Testing Laboratory	FRI, Dehradun	Uttar Khand
13	Seed Testing Laboratory	Delhi Road Meerut	Uttar Pradesh
14	Seed Testing Laboratory	Rampur Garden, Bareilly	Uttar Pradesh
15	Seed Testing Laboratory	Civil Lines, Mathura	Uttar Pradesh
16	Seed Testing Laboratory	GATDC, Jhansi	Uttar Pradesh
17	Seed Testing Laboratory	GATDC, Hardoi	Uttar Pradesh
18	Seed Testing Laboratory	GATDC, Varanasi	Uttar Pradesh
19	Seed Testing Laboratory	GATDC, Barabanki	Uttar Pradesh
20	Seed Testing Laboratory	GATDC, Azamgarh	Uttar Pradesh
21	Seed Testing Laboratory	A-264 Indira Nagar, Lucknow	Uttar Pradesh
22	Seed Testing Laboratory	Agri. Uni., Faizabad	Uttar Pradesh
23	State Seed Testing Laboratory	Gwalior	Madhya Pradesh
24	Seed Testing Laboratory	Krishi Nagar, Jabalpur	Madhya Pradesh
25	Seed Testing Laboratory	Agri. Uni., Nagpur	Maharashtra
26	Seed Testing Laboratory	Agri. College, Pune	Maharashtra
27	Seed Testing Laboratory	MAU, Parbhani	Maharashtra
28	Seed Testing Laboratory	Agri. Dept. Aurangabad	Maharashtra
29	Seed Testing Laboratory	MPKV, Rahuri	Maharashtra
30	Seed Testing Laboratory	PDKV, Akola	Maharashtra
31	State Seed Testing Laboratory	Agri. Dept. Dholi	Bihar
32	Seed Testing Laboratory	Sheikhpura, Patna	Bihar
33	Seed Testing Laboratory	RAU, Sabour	Bihar
34	Seed Testing Laboratory	BAU, Ranchi	Bihar

Sl.No.	*Name of Laboratory*	*Location/Address*	*State*
35	Seed Testing Laboratory	Ulubari, Gauhati	Assam
36	Seed Testing Laboratory	Shillong	Meghalaya
37	Seed Testing Laboratory	Malda	West Bengal
38	Seed Testing Laboratory	Kolkata	West Bengal
39	Seed Testing Laboratory	Agri Farm, Burdwan	West Bengal
40	Seed Testing Laboratory	Agri Dept., Ranipool	Sikkim
41	Central Seed Testing Laboratory	IARI, New Delhi	Delhi
42	Quality Control Laboratory	Pusa Campus, New Delhi	Delhi
43	Seed Testing Laboratory	Khyber Pass, New Delhi	Delhi
44	Seed Testing Laboratory	Agri. Uni., Junagarh	Gujarat
45	Seed Testing Laboratory	Agri. Dept., Navsari	Gujarat
46	Seed Testing Laboratory	Gandhi Nagar	Gujarat
47	Seed Testing Laboratory	Rajendranagar, Hyderabad	Telangana
48	Seed Testing Laboratory	Tandepalligudam	Andhra Pradesh
49	Seed Testing Laboratory	Lalaguda, Secunderabad	Telangana
50	Seed Testing Laboratory	Agri. Dept. Cuddapah	Andhra Pradesh
51	Seed Testing Laboratory	Hebbal, Bangalore	Karnataka
52	Seed Testing Laboratory	Lalbagh, Bangalore	Karnataka
53	Seed Testing Laboratory	Dharwad	Karnataka
54	Seed Testing Laboratory	Agri Dept. Bhubneshwar	Odisha
55	Seed Testing Laboratory	Agri. Dept., Panji	Goa
56	Seed Testing Laboratory	Indramanipuram, Coimbatore	Tamil Nadu
57	Seed Testing Laboratory	Tirunagar, Madurai	Tamil nadu
58	Seed Testing Laboratory	Vyalogam, Kundumian Malai	Tamil Nadu
59	Seed Testing Laboratory	ARS Pattambi, Palghat	Kerala
60	Seed Testing Laboratory	Kalar Kode, Alleppy	Kerala
61	Seed Testing Laboratory	Agri Dept. Pondicherry	Pondicherry

*****

Appendix 5
Chromosome Number in some Crop Plants

Sl.No.	*Common Name*	*Scientific Name*	*2n Number*	*n Number*
(a)	Monocots			
I	Cereals			
1	Bread Wheat	*Triticum aestivum*	42	11
2	Macroni wheat	*Triticum duram*	28	14
3	Triticale	*Triticale*	56	07
4	Rye	*Secale cereale*	14	07
5	Barley	*Hordeum vulgare*	14	07
6	Rice	*Oriza sativa*	24	12
7	Maize	*Zea mays*	20	10
8	Sorghum	*Sorghum bicolour*	20	10
9	Pearl millet	*Penisetum americanum*	14	07
(b)	Dicots			
II	Pulses			
10	Urd bean	*Vigna mungo*	22	11
11	Mung bean	*Vigna radiata*	22	11
12	Red gram	*Cajanus cajan*	22	11
13	Cow pea	*Vigna anguiculata*	22	11
14	Moth bean	*Vigna aconotifolia*	22	11
15	French bean	*Phaseolus vulgaris*	22	11
16	Indian bean [Sem]	*Lablab purpureus*	22	11
17	Garden Pea	*Pisum sativum*	14	07
18	Chick pea	*Cicer aritinum*	16	08
19	Lentil	*Lens culinaris*	14	07
20	Khesari	*Lathyrus sativus*	14	07
21	Cluster bean	*Cymopsis tetragonoloba*	14	07
22	Soybean	*Glycine max*	40	20
III	Oilseeds			
23	Ground nut	*Arachis hypogea*	40	20
24	Mustard	*Brassica juncea*	36	18
25	Banarsi Rai	*Brassica nigra*	16	08
26	Sesame	*Sesamum indicum*	26,52	13, 26
27	Sunflower	*Helianthus annus*	34	17
28	Safflower	*Cartamus tinctorius*	24	12
29	Castor	*Ricinus communis*	20,40	10,20
30	Linseed	*Linum usitatissimum*	30,32	15, 16

Sl.No.	Common Name	Scientific Name	2n Number	n Number
IV	Vegetables			
31	Potato	*Solanum tuberosum*	48	24
32	Cabbage	*Brassica oleracea*	18	09
33	Cucumber	*Cucumis sativus*	14,28	07, 14
34	Ridge gourd	*Luffa acutangula*	26	13
35	Sponge gourd	*Luffa cylindrica*	26	13
36	Tomato	*Lycopersicon esculentum*	24	12
37	Okra	*Abelmoschus esculetus*	72	36
38	Brinjal	*Solanum melogena*	24,48	12,24
39	Carrot	*Daucus carrota*	18	09
40	Radish	*Raphanus sativus*	18	09
41	Chilli	*Capsicum annum*	24	12
42	Onion	*Allium cepa*	16	08
43	Garlic	*Allium sativum*	16	08
V	Fibre Crops			
44	Upland Cotton	*Gossypium hirsutum*	52	26
45	Egyptian cotton	*Gossypium barbadennse*	52	26
46	Tree cotton	*Gossypium arboreum*	26	13
47	Levant cotton	*Gossypium herbaceum*	26	13
48	Jute	*Corchorus sp.*	14	07
49	Sunhemp	*Crotolaria juncea*	16	08
50	Manila hemp	*Musa textilis*	20	10
VI	Commercial Crops			
51	Tobacco	*Nicotiana tabacum*	48	24
52	Sugarcane	*Sacharum officinarum*	80	40

*****

Appendix 6
List of State Seed Corporations

Sl.No.	*State*	*State Seed Corporation*
1	Andhra Pradesh	Andhra Pradesh State Seeds Development Corpn. Ltd. 5-10-193 (2nd Floor) HACA Bhavan, Opp. Public Gardens Hyderabad – 500 004.
2	Karnataka	Karnataka State Seeds Corporation Ltd. Beej Bhavan, Bellary Road, Hebbal, Bangalore – 560 024.
3	Rajasthan	Rajasthan State Seeds Corporation Ltd. Pant Krishi Bhawan, B.D. Road, Jaipur O.P. Saini, IAS Managing.
4	Punjab	Punjab State Seeds Corporation Ltd. S.C.O. Nos. 835-836, Sector 22-A, Chandigarh.
5	Gujarat	Gujarat State Seeds Corporation Ltd. Beej Bhavan, Sector 10 -A, Gandhinagar – 382 010, Gujarat.
6	Haryana	Haryana Seeds Development Corporation Ltd. Bays No.308, Sector - 2 Panchkula – 134 112, Haryana
7	Uttaranchal	Uttarakhand Seeds and Tarai Development Corpn. Ltd. Pantnagar, P.O.: Haldi, Distt.: U.S. Nagar – 263 139, Uttarakhand
8	West Bengal	West Bengal State Seeds Corp. Ltd. 4, Gangadhar babu Lane (5th floor), Kolkata – 700 012, W.B.
9	Madhya Pradesh	MP Seeds and Farms Development Corpn.Ltd. E-1/88, Arera Colony, Bhopal, M.P.
10	Odisha	Orissa State Seeds Corporation Ltd. Asha Nivas, Lewis Road, Bhubneswar, Odisha.
11	Bihar	Bihar Rajya Beej Nigam Ltd. Indira Bhawan, 2nd Floor, Patna – 577 583, Bihar.
12	Maharashtra	Maharashtra State Seed Corporation Ltd. Mahabeej Bhavan, Krishi Nagar, Akola – 444 104, Maharashtra.
13	Assam	Assam State Seeds Corporation Ltd., Madhura Nagar, Dispur, Assam.
14	Uttar Pradesh	U.P. Seeds Development Corporation C-973/74B Faizabad Road, Mahanagar, Lucknow – 226 006, U.P.

*****

Appendix 7
List of State Seed Certification Agencies

Sl.No.	*State*	*Seed Certification Agency*
1	Andhra Pradesh	Andhra Pradesh State Seed Certification Agency, Hyderabad.
2	Assam	Assam State Seed Certification agency, Gauhati.
3	Bihar	Bihar State Seed Certification agency, Patna.
4	Gujarat	Gujarat State Seed Certification agency, Ahmedabad.
5	Haryana	Haryana State Seed Certification agency, Chandigarh.
6	Himachal Pradesh	Himachal Pradesh State Seed Certification agency, Shimla.
7	Jammu and Kashmir	Seed Certification wing, Srinagar.
8	Karnataka	Karnataka State Seed Certification agency, Bangalore.
9	Kerala	Department of Seed Certification, Trivandrum.
10	Madhya Pradesh	M.P. State Seed Certification agency, Bhopal.
11	Maharashtra	Maharashtra State Seed Certification agency, Pune.
12	Odisha	Orissa State Seed Certification agency, Bhubneshwar.
13	Punjab	Punjab State Seed Certification agency, Chandigarh.
14	Rajasthan	Rajasthan State Seed Certification agency, Jaipur.
15	Sikkim	Seed Certification wing,
16	Tamil Nadu	Department of Seed Certification, Coimbatore.
17	Uttar Pradesh	Uttar Pradesh State Seed Certification agency, Lucknow.
18	West Bengal	West Bengal State Seed Certification agency, Kolkata.
	Union Territory	
19	Delhi	Seed Certification Unit, Delhi.
20	Puducherry	Seed Certification wing, Puducherry.

*****

Glossary

PART I: INTRODUCTORY TOPICS

Chapter 1: Modes of Reproduction in Crop Plants

Adventive Embryony: Development of embryo from the diploid cells of ovule lying outside the embryosac belonging to either nucellus or integument.

Allogamy: Development of seed by cross pollination.

Anthesis: The process of dehiscence of anthers, the period of pollen distribution.

Apomixis: Development of seed without sexual fusion (fertilization).

Apogamy: Development of embryo either from synergids or antipodal cells of embryosac.

Apospory: Development of another embryosac without reduction from the cell of ovule outside the embryosac and then development of embryo directly from diploid egg cell.

Artificial Vegetative Reproduction: It is achieved by artificial means.

Asexual Reproduction: Multiplication of plants without the fusion of male and female gametes. It is of two type, *viz.* : vegetative reproduction and apomixis.

Autogamy: Development of seed by self-pollination.

Bisexuality: Presence of male and female organs in the same flower.

Chasmogamy: Fertilization after opening of flower.

Cleistogamy: Completion of pollination and fertilization in unopened flower bud.

Dichogamy: Maturation of anthers and stigma of a flower at different time.

Dioecy: Existence of male and female flowers on different plants.

Diploid Parthenogenesis: Development of embryo from diploid egg cell.

Facultative Apomixis: Apomixis in which sexual reproduction occurs in addition to apomixis.

Generative Apospory: Development of new embryosac may from archesporium.
Somatic Apospory: Development of new embryosac from nucellus or integument.

Herkogamy: Hindrance to self-pollination due to presence of hyline membrane around anther.

Heterostyly: Different lengths of styles and filaments in a flower.

Homogamy: Maturation of anthers and stigma of a flower at the same time.

Monoecy: Separate existence of male and female flowers on the same plant in the same inflorescence or different inflorescence.

Natural Vegetative Reproduction: It occurs in nature.

Non-Recurrent Apomixis: When apomictic plants are not obtained from one generation to another.

Obligate Apomixix: Apomixis in which reproduction occurs only by apomixis.

Parthenogenesis: Development of embryo from the egg cell without fertilization.
Haploid Parthenogenesis: Development of embryo from haploid egg cell.

Recurrent Apomixis: Occurrence of apomixis from one generation to another.

Reproduction: The process by which living organisms give rise to the offspring of similar kind (species).

Sexual Reproduction: Multiplication of plants through fertilized embryos.

Vegetative Reproduction: Multiplication of plants by means of various vegetative plant parts such as stem cuttings and root cuttings.

Chapter 2: Types of Improved Seed

Breeder seed: The progeny of either nucleus seed or breeder seed produced under the direct supervision of original or sponsoring plant breeder.

Certified seed: The progeny of either foundation or certified seed

Foundation seed: The progeny of breeder seed produced by NSC or SSC.

Improved Seed: The seed of a release and popular variety produced by scientific method; also called quality seed.

Nucleus seed: The initial seed of a released and notified variety which is always limited in quantity.

Registered seed: The progeny of either foundation or registered seed.This category is omitted in India.

Variety: A genotype which has been released for commercial cultivation either by State Variety Release Committee or Central Variety Release Committee.

Chapter 3: Development of Inbred Lines

Double Cross Hybrid: The hybrid progeny from a cross between two single crosses, *i.e.* (A x B) x (C x D).

Double Top-cross: A cross between a single cross and an open pollinated variety, *viz.* (A x B) x OP variety.

Hybrid Variety: The F_1 population that is used for commercial cultivation.

Hybrid: The progeny of a cross between genetically dissimilar parents.

Inbred: In cross pollinated species, a true breeding line that is obtained by continuous inbreeding

Interspecific Cross: A cross between two different species of the same genus.

Intraspecific Cross: A cross between two varieties or genotypes of the same species; also called inter-varietal cross.

Multiple Cross: A cross involving more than four inbred lines; also known as composite cross.

Poly cross: Open pollination of a group of selected genotypes in isolation from other compatible genotypes to promote random mating among selected genotypes.

Single Cross Hybrid: The hybrid progeny from a cross between two inbreds or varieties, *viz.* (A x B).

Three way Cross Hybrid: The hybrid progeny between a single cross and an inbred *viz.* (A x B) x C.

Top Cross: A cross between an inbred line and an open pollinated variety; also known as inbred variety cross.

Chapter 4: Development of Hybrids Using A, B and R Lines

Double Cross Hybrid: The hybrid progeny from a cross between two single crosses, *i.e.* (A x B) x (C x D).

Double Top-cross: A cross between a single cross and an open pollinated variety, *viz.* (A x B) x OP variety.

Hybrid Variety: The F_1 population that is used for commercial cultivation.

Hybrid: The progeny of a cross between genetically dissimilar parents.

Inbred: In cross pollinated species, a true breeding line obtained by continuous inbreeding

Interspecific Cross: A cross between two different species of the same genus.

Intraspecific Cross: A cross between two varieties or genotypes of the same species; also called inter-varietal cross.

Multiple Cross: A cross involving more than four inbred lines; also known as composite cross.

Poly cross: Open pollination of a group of selected genotypes in isolation from other compatible genotypes to promote random mating among selected genotypes.

Single Cross Hybrid: The hybrid progeny from a cross between two inbred lines or varieties, *viz.* (A x B).

Three way Cross Hybrid: The hybrid progeny between a single cross and an inbred, *viz.* (A x B) x C.

Top Cross: A cross between an inbred line and an open pollinated variety; also known as inbred variety cross.

Chapter 5: Variety Release Procedure

Clonal variety: A variety which is developed by clonal selection, relevant to asexually propagated species.

Hybrid variety: A F_1 population which is used for commercial cultivation.

Multi-line variety: A variety of self-pollinated crops which is developed by mixing seed of several isogenic lines, related lines or unrelated lines in equal quantity.

Mutant variety: A variety which is developed by mutation breeding.

Pedigree bred variety: A variety which developed by pedigree breeding method.

Pure line variety: A variety which is developed by pure line selection.

Strain: A group of genotypes that are similar in phenotype. After release a strain is called variety.

Transgenic variety: A variety which is developed by the technique of genetic engineering. It differs in one character from the parent variety.

Variety: A strain which is released for commercial cultivation by a variety release committee.

PART II: SEED PRODUCTION, CERTIFICATION AND TESTING

Chapter 6: Principles of Quality Seed Production

Grow-out Test: A test which is conducted to assess the genetic purity of a variety during seed production.

Isolation Distance: The separation of the field of a variety from that of another variety of the same crop by a minimum prescribed distance to avoid contamination.

Rogueing: The process of removal of the off types plants from the field of an improved variety to avoid contamination.

Seed certification: A legal system which ensures availability of high quality seed of improved variety to the farmers.

Volunteer plants: The plants which grow in the field from previous crops.

Chapter 7: Hybrid Seed Production: Field Crops

Double Cross Hybrid: The hybrid progeny from a cross between two single crosses, *i.e.* (A x B) x (C x D).

Hybrid Variety: The F_1 population that is used for commercial cultivation.

Hybrid: The progeny of a cross between genetically dissimilar parents.

Isolation Distance: The separation of the field of a variety from that of another variety of the same crop by a minimum prescribed distance to avoid contamination.

Rogueing: The process of removal of the off types plants from the field of an improved variety to avoid contamination.

Single Cross Hybrid: The hybrid progeny from a cross between two inbreds or varieties, *viz.* (A x B).

Three way Cross Hybrid: The hybrid progeny between a single cross and an inbred, *viz.* (A x B) x C.

Chapter 8: Hybrid Seed Production: Vegetable Crops

Double Cross Hybrid: The hybrid progeny from a cross between two single crossesm, *i.e.* (A x B) x (C x D).

Hybrid Variety: The F_1 population that is used for commercial cultivation.

Hybrid: The progeny of a cross between genetically dissimilar parents.

Isolation Distance: The separation of the field of a variety from that of another variety of the same crop by a minimum prescribed distance to avoid contamination.

Rogueing: The process of removal of the off types plants from the field of an improved variety to avoid contamination.

Single Cross Hybrid: The hybrid progeny from a cross between two inbreds or varieties, *viz.* (A x B).

Three way Cross Hybrid: The hybrid progeny between a single cross and an inbred, *viz.* (A x B) x C.

Chapter 9: Seed Certification

Certified seed: The seed produced from foundation, registered or certified seeds.

Grow-Out Test: The test which is carried out to evaluate the seeds for their genuineness to species or varieties or seed borne infection.

Seed Certification Agency: A government organization which is authorized for seed certification. It may have its own seed testing laboratory or it may get its seed samples tested through seed testing laboratories.

Seed Certification Measures: Various measures which are adopted to ensure quality seed production such as verification of seed source, field inspections, sample inspection, bulk inspection, control plot test and grow-out test.

Seed Certification Standards: The minimum standards of isolation distance, genetic purity and germination required for the certification of seeds as prescribed by the certification agencies.

Seed Certification: It is a legal system which ensures availability of high quality seed of improved variety to the farmers.

Chapter 10: Testing Seed Quality

Abnormal Seedlings: Defective seedlings that lack either cotyledons or have stunted root and shoot or their essential structures are so much decayed that they cannot develop into normal plants.

Composite Sample: The seed sample which is formed by mixing all the primary samples of a seed lot.

Dead Seeds: These are nonviable seeds.

Fresh Ungerminated Seeds: These are viable seeds but do not germinate and remain fresh in germination test.

Hard Seeds: The seeds which do not absorb water. Such seed are common in Fabaceae and Malvaceae families. Seed coats of such seeds are impermeable to water.

Inert Matter: Broken or damages seeds, leaf bits, straw, soil particles, stones, *etc.*

Normal Seedlings: Seedlings with normal growth of root and shoot.

Other Seeds: Seeds of other variety of a crop, other crop seeds and weed seeds.

Primary Sample: The small portion of seed drawn from the large seed lot. Several samples are taken from the seed lot to represent the composition of the seed lot.

Pure Seeds: The portion of the seed that belongs to the variety under testing.

Sampling Authority: An officer or inspector of State Seed Certification Agency and authorized to draw seed samples from the seed lots.

Sampling: The process of taking a small portion of seed from the large seed lot.

Seed Testing: The laboratory analysis of a seed lot to determine its quality in terms of physical purity, seed health, germination and seed moisture content

Seed Viability: The capacity of seeds to germinate under favorable conditions in the absence of dormancy.

Seed vigour: The general ability of a seed lot to germinate normally over a range of adverse conditions.

Submitted Sample:The seed sample which is submitted to the seed testing laboratory for testing.

Working Sample: The seed sample which is drawn from the submitted seed sample to carry out the seed testing work.

Chapter 11: Maintenance Breeding

Maintenance breeding: An area of plant breeding which deals with principles and methods of breeder seed production; also known as varietal maintenance technology.

Model Bulk Selection: Selecting large number of single plants [300-500] from the breeder seed plot of a variety, examining these plants under laboratory conditions for the characteristics of the variety recorded at the time of its release and selecting those plants which conform to the characteristics of a variety and their seed is bulked to grow nest generation.

Negative Mass Selection: Removal of undesirable plants [off types] from the field of a variety and mixing the seed of remaining plants to grow the next generation.

Plant to Row Method: This method of maintenance breeding is based on progeny test. It is also known as progeny row selection. Progeny of single plant selections are planted, evaluated and progeny with true to the type are selected and their seed is mixed to grow the nest generation.

Positive Mass Selection: Selection of desirable plants [true to type] from the field of a variety and mixing the seed of individual selected plants to grow the next generation.

PART III: MISCELLANEOUS TOPICS

Chapter 12: Principles of DUS Testing

Candidate variety: A variety to be registered under Plant Variety Protection Act. **Reference Variety:** All released and notified extant varieties of common knowledge which are in seed production chain.

Distinctiveness: The new variety must be clearly distinguishable by one or more characters from any other variety whose existence is a matter of common knowledge at the time when the protection is applied for.

DUS Testing: Evaluation of plant varieties in terms of distinctiveness, uniformity and stability.

Example Variety: A variety that is used for comparison for a particular character. **Farmers' Variety**: A variety that has been developed by a farmer and used for commercial cultivation for several years.

Extant Variety: All released, notified and unprotected varieties.

Farmers' Rights: Legal rights provided to the farmers to save, use, sow, replant, exchange, share or sell his farm produce including seed of a variety protected under Plant Variety Protection Act.

Novelty: It refers to newness of a variety. The variety should be new one and should not have been commercially cultivated for more than one year before granting protection under PVP Act.

Plant Breeders' Rights: Legal rights granted to the plant breeder for a new plant variety. It provides rights for commercial seed production, marketing, export and import, authorization and to prevent infringement.

Stability: The variety is deemed to be stable if its relevant characteristics remain unchanged after repeated propagation.

Uniformity: The variety should be sufficiently uniform in its relevant characteristics. Relevant characteristics include all those used as a basis for distinctiveness or included in the variety description established at the date of grant of protection of that variety.

Chapter 13: PPV and FR Act 2001

Candidate Variety: A variety that has to be protected.

Distinctiveness: The variety should be clearly distinguishable from previously available varieties at least for one characteristic.

Example Variety: A variety that is used for comparison of a particular character.

Extant Variety: All released and notified varieties which have not been protected.

Farmers' Variety: A variety that has been developed by a farmer and used for commercial cultivation for several years.

Novelty: Newness of a variety. A variety which has not been grown for more than one year prior the application for registration.

Reference variety: All released and notified extant varieties that are under seed production chain.

Stability: The same performance of a variety after repeated reproduction or propagation.

Uniformity: The same type of population of a variety.

Chapter 14: Intellectual Property Rights

Copyright: The legal right granted to an author, composer, playwright, publisher, or distributor to exclusive publication, production, sale, or distribution of a literary, musical, dramatic, or artistic work.

Database Rights: An intellectual property right given to a computer database.

Domain Name: A unique name that identifies an Internet site.

Farmers' Rights: Legal rights provided to farmers to save, use, sow, replant, exchange, share or sell his farm produce including seed of a variety protected under Plant Variety Protection Act

Immovable Property: Fixed types of properties which cannot move from one place to other. Such properties include land, buildings [house, flat, bungalow, farm house] and gardens.

Industrial Design: An industrial design - or simply a design - is the ornamental or aesthetic aspect of an article produced by industry or handicraft.

Intellectual Property: Creations of the mind such as inventions, literary and artistic works, and symbols, names, images, and designs used on commercial scale.

Mask work Right: A mask work is a two or three-dimensional layout of an integrated circuit.

Moral Rights: Moral rights are a special extension to copyright that gives the originator of a copyrighted work rights over its use.

Movable Property: The property which can be easily shifted from one place to other. Such property includes animals, farm machines, furniture, fixtures, gold, silver, diamond, and money.

Patent: An official document which grants sole rights to the inventor for manufacturing and marketing his product/process/invention to derive benefits.

Plant Breeder' Rights: Legal rights granted to the breeder of a new variety of plant.

Sui-generis: Things of their own kind or things with unique characteristics.

Supplementary Protection Certificate: An extension of the term given to pharmaceutical or plant protection patent.

Trade Name: The name under which a business trades for commercial purposes.

Trade Secret: A formula or process or device used in business that is not published or divulged which gives an advantage over competitors.

Trademark: A distinctive sign of some kind which is used by an individual, business organization or other legal entity to uniquely identify the source of its products and/or services to consumers, and to distinguish its products or services from those of other entities.

Traditional Knowledge: The knowledge, innovations and practices of indigenous people and local communities.

Utility Model: An exclusive right granted for an invention, which allows the right holder to prevent others from commercially using the protected invention, without his authorization, for a limited period of time.

Chapter15: Alternate Strategies for Line and Cultivar Development [Haploids, Somatic hybridization, Tissue Culture and Biotechnology]

(*i*) *HAPLOIDY*

Allohaploid: Haploid that develops from allopolyploid species.

Aneuhaploid: Haploid that develops from an aneuploid species.

Autohaploid: Haploid that develops from autoployploid species.

Euhaploid: Haploid which develop from a euploid species.

Haploid: An individual with half of somatic chromosome number.

Monohaploid: Haploid which develops from a normal diploid species.

Monoploids: Individuals with basic chromosome number of a species. Such condition is called monoploidy.

Polyhaploid: Haploid that develops from a polyploid species.

(*ii*) *SOMATIC HYBRIDIZATION*

Asymmetrical somatic hybrids: Somatic hybrids that contain complete somatic complement of one species and only part of the somatic complements of other species.

Cybrids: Somatic hybrids with normal protoplast of one species and nucleusless protoplast of other species.

Cytoplast: A protoplast either without nucleus or with inactive nucleus.

Heterokaryons: Hybrid cell involving protoplasts of two different species.

Homokaryons: Hybrid cell involving protoplasts of the same species.

Inter-generic somatic hybrids: Hybrids obtained by protoplast fusion between two different genera of the same family.

Inter-specific somatic hybrids: Hybrids obtained by protoplast fusion between two different species of the same genus.

Inter-tribal somatic hybrids: Hybrids obtained through protoplast fusion between plants of two different families.

Protoplast: Naked cells or cells without cell wall.

Somatic hybridization: Crossing of crop plants through fusion of somatic cells.

Somatic Hybrids: Hybrids obtained by somatic hybridization.

Symmetrical somatic hybrids: Somatic hybrids that contain all chromosomes of both the species involved in the fusion of protoplast.

(*iii*) *TISSUE CULTURE*

Callus: An unorganized mass of differentiated plant cells.

Clonal propagation: Asexual multiplication of plants from a single individual or explant.

Culture: A plant growing *in vitro* in a sterile environment.

Doubled Haploid: Plants derived from a haploid after doubling chromosome number colchicine treatment to form homozygous diploids.

Embryo Rescue: *In vitro* culture technique used to rescue inherently weak, immature or hybrid embryos to prevent degeneration.

***Ex vitro*:** Organisms removed from tissue culture and transplanted; generally plants to soil or potting mixture.

Explant: An excised piece or part of a plant used to initiate a tissue culture.

Micro-propagation: Multiplication of plants from vegetative parts by using tissue culture nutrient medium.

Protoplast: Cells without cell wall, usually removed by digestion with enzymes.

Somaclonal Variation: Genetic variation among progeny of plants regenerated from somatic cells cultured *in vitro*.

Somatic embryos: Non-zygotic bipolar embryo-like structures obtained from somatic cells.

Tissue Culture: Growth of tissues or cells in an artificial medium.

Totipotency: Capacity of plant cells to regenerate whole plants when cultured on media.

Transgenic: Plants that have a piece of foreign DNA.

(iv) *TRANSGENIC TECHNOLOGY*

Agrobacterium tumifaciens: A soil-borne bacterium which is widely used for development of transgenic plants in different crop plants. It is a biological method of foreign gene transfer into host cells.

Biotechnology: Cellular and bio-molecular processes to develop technologies and products that can be commercialized.

Gene Cloning: The process of making several identical copies of a gene in genetic engineering.

Genetic Engineering: Isolation, introduction and expression of foreign DNA in plants and animals.

Genetic Transformation: The process of transfer, integration and expression of transgene (foreign gene) in an organism.

Micro-injection: A mechanical method of gene transfer in which foreign DNA is delivered into the host cells by microscopic needles also known as micro-injection.

Particle Bombardment: A method of foreign gene transfer which involves high velocity metal particles (usually gold or tungsten) or metal particles mediated method of genetic transformation. It is also known as biolistic micro projectile bombardment, micro ballistics and particle acceleration.

Plant Biotechnology: Combined study of plant tissue culture and genetic engineering or recombinant DNA technology.

Plasmid: An extra-chromosomal genetic element which is found in the bacteral cell, can replicate independently of chromosomal DNA and is not essential for normal growth and development of bacterium.

Transformants: Cells or tissues of an organism which have udergone genetic transformation; resultants of genetic transformation.

Transgenes: Foreign genes or modified genes of the same species which are used for the development of transgenic individuals. Such gene may be from the same species (in modified form), related wild species, unrelated species and microbes (bacteria, fungi and viruses).

Transgenic Breeding: Genetic improvement of crop plants, domestic animals and useful micro-organisms through biotechnology.

Transgenic: Genetically engineered organisms or organisms developed by the techniques of genetic engineering. It may be a plant, an animal or micro-organism.

*****

References

1. **ALLARD, R. W.** 1960. Principles of Plant Breeding. John Wiley and Sons Inc. New York.
2. **ALLARD, R. W.** 1999. Principles of Plant Breeding 2nd edition. John Wiley and Sons Inc. New York.
3. **BOROJEVIC, S.** 1990. Principles and Methods of Plant Breeding. Elsevier, Amsterdam.
4. **BRIGGS, F. N. AND KNOWLES, P. F.** 1967. Introduction to Plant Breeding. Reinhold, New York.
5. **CHOPRA, V. L. (ED.)** 2000. Plant Breeding: Theory and Practice 2nd edition, Oxford and IBH Publishing Company, Pvt. Ltd. New Delhi.
6. **FEHR, W. R.** 1987. Principles of Cultivar Development Volume 1. Theory and Techniques. Macmillan Publishing Company, New York.
7. **FREY, K. J.** (ed.) 1966. Plant Breeding. Iowa State University, Press, Ames, Iowa, USA.
8. **FREY, K. J.** (ed.) 1981. Plant Breeding II. Iowa State University, Press, Ames, Iowa, USA.
9. **HARTEN, A. V.** 1998. Mutation Breeding: Theory and Practical Applications. Cambridge Uni. UK.
10. **KUCKKUCK, H., KOBADE, G. AND WENGEL G.** 1993. Fundamentals of Plant Breeding. Narosa Publishing House, New Delhi.
11. **MAYO, O. 1984**. The Theory of Plant Breeding. Clarendon, Oxford.
12. **POEHLMAN, J. M.** 1987. Breeding Field Crops 3rd edn.AVI Publishing Co. Inc. West Port, Connecticut, USA.
13. **POEHLMAN, J. M. AND BORTHAKUR, D. N.** 1969. Breeding Asian Field Crops. Oxford and IBH Publishing Company, New Delhi.

14. **PHUNDAN SINGH** 2018. Essentials of Plant Breeding 7th edition, Kalyani Publishers, New Delhi.

15. **PHUNDAN SINGH** 2018. Plant Breeding: Molecular and New Approaches 4th edition, Kalyani Publishers, New Delhi.

16. **PHUNDAN SINGH** 2017. Fundamentals of Plant Breeding, Kalyani Publishers, New Delhi.

17. **PHUNDAN SINGH** 2018. Concepts in Plant Breeding, Brillion Publishing House, New Delhi.

18. **PHUNDAN SINGH** 2018. Introduction to Maintenance Plant Breeding, Daya Publishing House, New Delhi.

19. **PHUNDAN SINGH AND PRATIBHA BISEN** 2020. Commercial Plant Breeding, Daya Publishing House, New Delhi.

20. **PHUNDAN SINGH** 2019. Elements of Genetics 5th edition, Kalyani Publishers, New Delhi.

21. **PHUNDAN SINGH** 2020. Genetics 4th edition, Kalyani Publishers, New Delhi.

22. **PHUNDAN SINGH** 2012. Cotton Breeding 3rd edition, Kalyani Publishers, New Delhi.

23. **PHUNDAN SINGH AND SANJEEV SINGH** 2010. Breeding Hybrid Cotton 3rd edn. Kalyani Publishers, New Delhi.

24. **SHARMA, J. R.** 1984. Principles and Practice of Plant Breeding. Tata McGraw-Hill Publishing Company Ltd. New Delhi.

25. **SIMMONDS, N. W.** 1979. Principles of Crop Improvement. Longman, London.

26. **SNEEP, J.AND HENDRIKSON, A. J. T.** (eds.) 1979. Plant Breeding Perspectives. Pudoc, Wageningen, Netherlands.

27. **VOSE, P. B. AND BLIXT, S. G.** (eds.) 1984. Crop Breeding A Contemporary Basis.Pergamon Press, London.

28. **WILLIAMS, W.** 1964. Genetic principles and Plant Breeding, Oxford, Blackwell.

*****

Books by the Senior Author

A Plant Breeding

1. Essentials of Plant Breeding
2. Introduction to Maintenance Plant Breeding
3. Plant Breeding: Molecular and New Approaches
4. Molecular Plant Breeding
5. Principles of Seed Technology
6. Seed Technology: At a Glance
7. Objective Seed Technology
8. Practical in Crop Breeding
9. Practical and Numerical in Plant Breeding
10. Numerical Problems in Plant Breeding and Genetics
11. Objective Science of Plant Breeding
12. Plant Breeding: At a Glance
13. Fundamentals of Plant Breeding [For UG Students]
14. Plant Breeding [For Under Graduate Students]
15. Genetics and Plant Breeding: Comparative Analysis
16. Genetics and Breeding of Stress Resistance in Crop Plants
17. Genetic Principles and Plant Breeding
18. Principles of Plant Breeding
19. Cotton Breeding
20. Heterosis Breeding in Cotton
21. Breeding Hybrid Cotton
22. Breeding Transgenic Bt. Cotton

23. Cotton Improvement in India
24. Glimpses of Cotton Breeding
25. Glossary of Plant Breeding and Genetics
26. Elements of Baby Corn
27. Breeding Crop Plants for Stress Resistance
28. Plant Breeding: Related Legislations

B. Genetics

29. Elements of Genetics
30. Genetics
31. Principles of Genetics [For UG Students]
32. Fundamentals of Genetics [For UG Students]
33. Plant Genetics
34. Cotton Genetics
35. Objective Genetics
36. Genetics: At a Glance
37. Objective Genetics and Plant Breeding
38. Objective Genetics and Plant Breeding [Hindi Edition]
39. Glossary cum Dictionary of Genetics and Plant Breeding
40. Molecular Genetics [Objective]
41. Molecular Genetics [At a Glance]
42. Molecular Genetics [Subjective]
43. Genetics and Man
44. Genetics Today

C. Quantitative Genetics

45. Biometrical Techniques in Plant Breeding
46. Application of Biometrical Techniques in Plant Breeding [Hindi Edn]
47. Objective Quantitative Genetics
48. Quantitative Genetics At a Glance
49. Quantitative Genetics

D. Plant Biotechnology

50. Introduction to Biotechnology
51. Plant Biotechnology
52. Principles of Plant Biotechnology
53. Plant Biotechnology: At a Glance

54. Objective Plant Biotechnology
55. Molecular Biology and Plant Biotechnology

E. Intellectual Property Rights

56. IPR and Plant Breeders' Rights [Subjective]
57. IPR and Plant Breeders' Rights [Objective]
58. IPR and Plant Breeders' Rights [At a Glance]
59. Introduction to Intellectual Property Rights
60. Intellectual Property Rights At A Glance
61. Intellectual Property Rights: Objective

F. Miscellaneous Books

62. A Guide to Competitive Examinations of Agriculture
63. Objective General Agriculture for Competitive Examinations
64. Shankar Kapas [Hybrid Cotton] in Hindi
65. Baby Corn Farming for Higher Income

*****

www.ingramcontent.com/pod-product-compliance
Ingram Content Group UK Ltd.
Pitfield, Milton Keynes, MK11 3LW, UK
UKHW021951270726
14060UKWH00002B/464